P9-BXV-539

ASTEROID

Earth Destroyer or
New Frontier ?

ASTEROID

Earth Destroyer or New Frontier ?

PATRICIA BARNES-SVARNEY

PLENUM PRESS • NEW YORK AND LONDON

Library of Congress Cataloging in Publication Data

Barnes-Svarney, Patricia L.
  Asteroid: earth destroyer or new frontier? / Patricia Barnes-Svarney.
    p.   cm.
  Includes bibliographical references and index.
  ISBN 0-306-45408-4
  1. Asteroids. 2. Asteroids—Collisions with Earth. I. Title
QB651.B36 1996   96-27009
523.4'4—dc20   CIP

ISBN 0-306-45408-4

Plenum Press is a Division of Plenum Publishing Corporation
233 Spring Street, New York, N.Y. 10013-1578

10 9 8 7 6 5 4 3 2 1

Printed in the United States of America

To my two (patient) siblings,
Karen and James,
and to all the scientists
who are keeping watch

PREFACE

When I was about 7 years old, my father set up his surveyor's transit instrument with its 22-power scope, turned the tube toward Jupiter, and held me up to the eyepiece. There, I saw the planet's bright disk, accompanied by four small "stars" that turned out to be the Galilean moons. I continued to stare until his arms got tired. As he lowered me to the ground, I made up my mind. I knew at that exact moment that I wanted to study the planets.

For years, my parents supported me through my adventures: They bought me my first telescope (a well-built 4-inch Newtonian reflector handmade by Don Yeier, who now owns Vernonscope in Candor, New York) and astronomy and geology books; drove me to local astronomy and geology clubs; stopped so I could collect rocks along the roadside during family trips (much to the chagrin of my sister and brother, who did not share my desire for muddy rocks); and on to college. My work stretched into time as an assistant curator of a planetarium and an observatory, and then as a geologist studying the best-known planet—the Earth.

Along the way, asteroids always seemed to take second place to my work. But no matter what I did, I still retained a chronic preoccupation with the smaller bodies of the solar system. I found out that the asteroids were the remnants of the early solar system, older than humans, dinosaurs, trilobites, or even the first vestiges of life on Earth. I learned that some of the asteroids were easier to visit via spacecraft than a voyage to the Moon and that they were often filled with precious metals there for the taking. And I eventually realized that we had a great deal in common with the asteroids: We were cousins, all of us made of star-stuff—born from the same elemental dust cloud that started our solar system.

I also found out there was a distinct possibility that we will be visited by one of these distant relatives in our lifetime. And when I realized it could really happen, I began to learn and write as much as possible about these elusive, potentially dangerous asteroid kin —which are yet a source of incredible resources.

My own opinions on asteroids and their impacts differ in some ways from those of many scientists, which is part and parcel of any field of science. Based on research conducted in the past two decades, many scientists, including myself, believe that asteroid impacts are possible and probable, and it will be the small asteroid that sneaks up on us that will cause the most damage. The major reason why such a strike could be devastating is our growing planetary population. As the human population increases along the coastal areas of the world, a tsunami from an impact in the oceans would endanger entire cities and towns. On land, the more people there are, the greater the possibility of a catastrophic tragedy if a large, or even relatively small, impacting body strikes one of the more populated sections of a continent.

Allow me to look back in time: A major reason astronomers watch the sky is because of the Earth's past. We now understand the awesome power of asteroids, since we have found impact crater scars to prove that the encounters took place. In addition, these small bodies have more than likely caused, or been the major catalyst for, mass extinctions in the past, and we do not want to be the next species up for extinction. We do not want other intelligent beings to look at our fossil record in millions of years, shake their heads in disbelief, and wonder why our species became extinct. We don't want them to wonder why we did not realize that we could have done something about one of the most potentially threatening processes in the solar system.

Many people believe that because scientists use the words *potential* or *probabilities* of a strike, we can ignore the asteroids. This is wrong because there are perils that scientists have brought up that we should address soon. And top scientists who have studied asteroids for years are warning us of possible danger.

Understandably, we are preoccupied with ailments and diseases that threaten us every day. Constantly bombarded with information about problems in politics, violence, civil wars, and ethnic cleansing, we tend to ignore the dangers that are not completely obvious to us (or those that the media hardly seems to cover). Out of *site*, out of mind, is the saying that best describes the majority of responses to asteroids. It is the same for people who ignore the potential for a major earthquake in California; or those who adhere to the standard Western response to medicine— ignoring the practice of preventative maintenance, and approaching our healthcare practitioners only when we become sick. If an asteroid were to strike the Earth and cause major damage and death, only then would the populace scream for cures and solutions, if, that is, we were lucky enough to be struck by a small impactor that did not destroy civilization altogether. By ignoring something that has the potential to destroy us, we are placing a bet that we may lose.

The impacts pelting Jupiter in 1994 gave us a taste of what is out there. The Earth may not have the gravitational pull or exposed surface area of a gas giant planet, but it is still part of the celestial target range. And although we may not see it, there may be a big bull's-eye on our planet.

CONTENTS

INTRODUCTION

Of all the damnable natural disasters are those
we can't control. But even more alarming are
those that also sneak up from behind.
(OVERHEARD AT ASTEROID CONFERENCE, 1991)

I do not mean to scare anyone, but there are asteroids out there with our name on them.

Allow me to qualify the above statement: First, we know that asteroids have struck our planet before—there is stark evidence in the form of large and small impact craters that dot the Earth's major landmasses and an occasional discovery in the oceans. Second, we have detected asteroids that orbit close to or cross the Earth's orbit. Which small body will strike, or when, is unknown, left in the hands of fate and orbital mechanics.

If you are still in doubt, picture this scenario: You are flying a standard cargo plane about 644 kilometers east of Tokyo, Japan. It is a routine flight, and the sky is somewhat clear with little wind. Suddenly, you observe an explosion nearby from the ocean, the blast quickly forming a giant mushroom cloud—spreading so fast that it rises from about 4000 to 18,000 meters in a matter of only 2 minutes. It is not your worst nightmare, a nuclear explosion, since there is no residual radiation; and though the oceans around Japan are geologically active, the shape and characteristic of the cloud rules out any possibility of underwater seismic or volcanic activity.

This scenario actually happened on April 9, 1984, to a Japanese cargo plane pilot. The conclusion was that the explosion was caused by an asteroid. It luckily missed the land, but managed to blast into the ocean, creating the awesome mushroom cloud. Scary, but true.

It is difficult to imagine a huge rock hurling through space on a collision course with Earth. To most people, rocks are nuisances found while planting a garden or trying to dig a foundation for a new home. Larger boulders serve as mere objects to rest on during a hike up a mountain pass. And, in general, if a rock from space—from the size of a thumbnail to those inhabiting the garden—were to plunge into the Earth's atmosphere, friction from our thick blanket of air would vaporize the object or break it into relatively harmless chunks of rock. In the majority of cases, what we would witness would be a meteor—a momentary trail of light in the nighttime sky.

Let's go from small to large in scale. Take one of the smallest asteroids found so far: An asteroid labeled 1991 BA came close to the Earth's orbit, passing just 170,000 kilometers from the Earth (less than half the distance to the Moon) in 1991. It is only about 9 meters across, or about the size of a small house. Toss this relatively small object at the Earth, and it would have the impact energy to destroy a small town.

Now take an asteroid that is 1 kilometer in diameter—just under 11 football fields end-to-end—and hurl it toward the Earth. The result would be a global catastrophe: As the impactor struck the surface, large amounts of dust would be injected high into the stratosphere, a layer of the Earth's atmosphere between 10 and 50 kilometers above the surface. The object would explode with energy approaching a million megatons of TNT, or equal to a 1-million-ton explosion (to compare, the largest reported warhead ever tested reached 57 megatons; the largest operational weapons yielded about 25 megatons).

The impact would have repercussions in almost every direction. There would be changes in global climate patterns (especially exacerbated by dropping temperatures from the dust in the atmosphere dimming sunlight), intense weather systems, and possible loss of food crops affected by the changes in climate patterns. Some scientists even warn that an impactor larger than a kilometer could cause an eventual breakdown of human society in a cascade effect—the loss of links within the food chain taking its toll on all living creatures on the planet.

Take a larger asteroid, say, the size of Vermont, and forget life on Earth altogether.

The purpose of this book is to put the asteroids of our solar system into perspective, explaining the history of asteroid discoveries, the details about the asteroids themselves, and even the potential that asteroids hold for our future in space and on Earth. It also presents the when, why, and how of an impact catastrophe, which can and will eventually take place on the Earth.

As a scientist and science writer, I have watched the odds of an impact as they have been hotly debated, tossed aside, gathered up again, and argued—the majority of the heated discussions having taken place during the past 20 years. Since the 1950s, scientists have realized that the Earth has been periodically struck by asteroids. But the scientific community had not widely accepted the idea that impacting objects could play a pivotal role as a major process on our planet.

The impetus for such revelations were the debates on planetary-wide species extinctions, in which some event or series of events caused the eradication of hundreds, if not thousands, of species over a short period of geologic time. Such debates began in earnest in 1980 after the publication of a paper by Luis Alvarez and others. They proposed that species extinctions between the Cretaceous and Tertiary (K-T) periods on the geologic time scale were caused by objects impacting the Earth. The paper set off an explosion of asteroidal studies in an attempt to understand the small bodies' place in solar system space and especially their effect on our planet—past, present, and future.

Certainly, asteroid impacts on Earth have lessened over the years, as impacting bodies have decreased in number since the beginnings of the solar system; but as we shall see, they have not stopped. Statistically speaking, an asteroid will collide with the Earth in the future, but like some cosmic bully pitching stones at our planet, sometimes the bully misses and sometimes the bully hits.

UPDATE ON CURRENT CONDITIONS

First, let us state the current condition of asteroidal affairs: There are no asteroids that we know of on a collision course with the Earth.

The qualification is necessary. No asteroids are heading for Earth, but the only population of asteroids we can speak of with any certainty are those we have already detected.

Most of these chunks of rock vary greatly in size, color, texture, and composition. We know that their histories are as diverse as they are: leftovers from the early formation of the solar system, broken offshoots of bigger parent asteroids, and no doubt comets that have lost their gassy glows. We know that 95 percent of all known asteroids circle the inner solar system planets just beyond the orbit of Mars and inside the orbit of Jupiter, at distances from 2.0 to 3.3 astronomical units from the Sun (the Earth is 1 astronomical unit from the Sun), a mere jump across the solar system pond in astronomical terms. We know that the total number of asteroids, from the asteroid main-belt to the rogue members that stray from the belt, comprise a few hundredths of a percent of the total matter orbiting the Sun.

We also know that so far, more than 6000 asteroids have already been discovered and, with additional observations, their orbits confirmed. Some of the top scientists in the field estimate asteroidal numbers to be closer to 10,000, perhaps more. Of the ones that come closest to the Earth (a subgroup called near-Earth asteroids), we have found about 250, with scientists estimating that the actual number averages around 1000. And it is these undiscovered bodies that scare us most.

Overall, most asteroids stay within the confines of a belt (called the main-belt of asteroids) between the orbits of Mars and Jupiter. But occasionally, an asteroid strays from the main-belt, pushed and pulled by some orbital gymnastics, changing its orbit to fit the new design physics has forced on it.

Such gymnastics are nothing new. Since the formation of the solar system, asteroids have never been static. For billions of years, these objects have collided with each other and have had their orbits gravitationally affected by planets, or perhaps even by passing stars or gas clouds. Unable to fight the forces of physics, many of the asteroids eventually roamed too far from the main-belt, either developing an eccentric orbit in the inner or outer solar system or catapulted out of the solar system altogether by the gravity of the larger planets or the Sun. Others developed orbits closer to the inner or outer planets, working their way to become planet-crossing asteroids (asteroids that cross the orbit of a solar system planet). Some planet-crossers cruise in their orbits, never coming near the planets, while others are yanked around to become planet-approaching asteroids. And if the movements are just right the asteroid's fate may be to eventually collide with a planet, including, of course, the Earth.

Other asteroids may actually be burned-out comets, their gases exhausted after multiple close runs past the Sun. Some have orbits that are strikingly similar to short-period comets, or those comets that spin around the Sun in periods under 200 Earth-years. Apparently, when these comets die out, their crusts harden over time. They then resemble an asteroid, and eventually their orbits will often come close to or cross over the Earth's orbit.

The dangers of asteroids is not a recent find. In 1941, Fletcher Watson—based on the discovery of the first Earth-approaching asteroids, Apollo, Adonis, and Hermes—estimated the rate of impacts on the Earth. Eight years later, Ralph Baldwin, in his book *The Face of the Moon*, noted:

> [S]ince the Moon has always been the companion of the Earth, the history of the former is only a phase of the history of the latter.... There is no assurance that these meteoritic impacts have all been restricted to the past. Indeed we have positive evidence that ... meteorites and asteroids still abound in space and occasionally come close to the Earth. The explosion that formed the crater Tycho [on the Moon] ... would, anywhere on Earth, be a horrifying thing, almost inconceivable in its monstrosity.

Since their first sightings two centuries ago, asteroids have had a bevy of names, including *minor planets, planetoids, meteoroids, moonlets, worldlets,* and *baby planets.* As of late, the name *asteroid* has carried its weight of interpretations, stretching from the inner solar system's planet-crossing bodies, the asteroid belt, and on to the outer reaches of the solar system, in which many icy, dark asteroidlike rocks exist. At least in size, asteroids, along with their cousins, the comets, are the babies of the solar system. Only twenty-six known asteroids have diameters greater than 200 kilometers, including the largest, Ceres, at about 1000 kilometers in diameter; the others include sizes less than 200 kilometers, down to house- and boulder-sized rocks (and there may be millions of the boulder-sized objects). To compare, the Earth's equatorial diameter is 12,756 kilometers; our satellite, the Moon, is 3476 kilometers in diameter; our smallest planet, Pluto, is 2445 kilometers in diameter; and the largest planet, Jupiter, has an equatorial diameter of 143,884 kilometers.

I am often asked why scientists are so concerned over such small objects of the solar system. With most asteroids limited to the asteroid belt, why should humans worry about misshapen objects 200 million kilometers away? Why should we even consider these small members of our solar system's neighborhood to be a threat?

But worries persist.

Hundreds, or maybe thousands, of asteroid chunks have worked their way into the inner solar system and set up orbits that take them close to the Earth, the so-called near-Earth asteroids. There are two weighty questions about these wanderers: How many are there, and where in the solar system are they? Scientists know now that small asteroids are notorious for showing up at the last minute in Earth-based telescopes, too dim to see before they are merely hundreds of thousand miles from our home!

The sizes of the errant objects closest to the Earth's orbit range from small boulders and the size of a house to as large as a major city. A little push or pull by the force of gravity, and Earth could experience an unwanted and unexpected close-up of such a wandering body. This has happened before: There are about 140 known surface impact craters (on every continent on Earth except Antarctica, where the craters may be hidden by the thick ice sheet) caused by collisions with space objects (asteroids and comets) in the past. No doubt many more have been erased or buried by the natural dynamics of our active planet—wind, water, plate tectonics, and mountain building. And as all seekers of near-Earth asteroids know, there may be an object out there with our name on it, one that is undetected until it strikes the Earth.

Amid the fears, more positive possibilities exist.

Asteroids often come close to the Earth, sometimes even closer than the Moon. What if we could lasso one of these near-Earth asteroids?

If such asteroids were available to us, we would have a potentially endless supply of resources with which to make our way into space. The strangely shaped rocks hold numerous necessary resources for living in space: Some hold water (in the form of water in permanently frozen ground) for fuel; some of the minerals in the carbonaceous (carbon-rich) asteroids are oxidized; and other asteroids could supply us with the rarer ores, such as gold, titanium, platinum, and associated valuable minerals necessary for building a space colony or to gather for the folks back home. And because most asteroids have small gravity wells (or low escape velocities), a rendezvous, and possible landings and takeoffs from the small bodies would be much easier than taking off from a major planet or even most of the satellites of the solar system. It is even feasible to put a specially designed space station on a larger asteroid or to turn a mined-out asteroid into living space.

Perhaps more than anything, the asteroids may tell us more about our solar system than all the data gathered by remote satellites or ground-based telescopes. The close bodies of rocks also hold the answers to the origin of our solar system. And they may even reveal more about the beginnings of life—including human life—on the Earth than any other source.

Chapter 1

MAJOR MINOR PLANETS

*Then I felt like some watcher of the skies when a
new planet swims into his ken.*
JOHN KEATS
EYEING THE HEAVENS

Vesta, the third-largest known asteroid in our solar system, is the only asteroid that can be seen with the naked eye. Vesta is smaller than Ceres and Pallas, the first- and second-largest asteroids, respectively, but its reflective surface allows it to be seen occasionally, without the aid of a telescope, in the nighttime sky. The rest of the known asteroids can only be seen with a scope or as a smudge on a photographic plate—their mere size living up to their "minor planets" label.

Because of their smallness, no astronomer prior to the invention of the telescope ever discovered an asteroid with the naked eye. Certainly, a sharper-eyed sky watcher could have spotted Vesta. But the speck in the sky, similar in appearance to the surrounding stars, was interpreted as just that—another faint star.

Instead, planets and stars monopolized the early writings and records in astronomy, with the occasional comets weighing in as the smallest bodies in the solar system. Even after the invention of the telescope, the planets, moons, comets, and stars dominated the astronomical limelight. It took two more centuries after the invention of the telescope for astronomers to actively seek a missing planet between Jupiter and Mars. The "planet" turned out to be a belt of small bodies—the first asteroid found with the usual luck that accompanies many scientific discoveries.

Before the invention of the telescope, one of the major scientific ideas that profoundly changed the way scientists perceived the solar system originated with Polish astronomer Nicholas Copernicus (1473–1543): The astronomer proposed a heliocentric view of the solar system, or that the Sun was the center of the solar system. (Actually, Aristarchus, in the third century B.C., first suggested that the Sun was at the center of the solar system, and movement of the heavens across the sky was caused by the Earth spinning on its axis. But Aristarchus' contemporaries objected to his ideas, pointing out that if the Earth did indeed rotate, unattached objects would fly off.)

For some 2000 years before Copernicus, ideas about the solar system were dominated by a geometric representation of the system, developed by Claudius Ptolemy (100?–170? A.D.), in which the Earth was at the center. Each planet moved uniformly in a small circle called an epicycle, with the Sun and Moon exempt from the epicycle motions. In turn, the center of each epicycle revolved uniformly around the Earth in a large circle called a deferent.

Fortunately for us, Copernicus had other ideas and the intelligence to record and interpret his observations. His father died when he was 10 years old, and he was raised by his uncle, a bishop, who provided Copernicus with an excellent education. After extensive studies at places such as the University of Cracow, Copernicus returned to his uncle's castle, working as a physician and diplomat, among other duties. With income from being elected a canon (because of his uncle's influence), Copernicus was finally, and fortuitously, able to spend time with his favorite study, astronomy. His studies went against Ptolemy's ideology, placing the Sun at the center of the solar system but still keeping the circular planetary orbits of the Ptolemaic system. The entire concept would soon catch on, leading astronomers to view the solar system in a very different way and to make discoveries that would reveal the system's true nature.

At this time in astronomy, there was no reason to suspect that Mars and Jupiter had company in their dances around the Sun. One of the first indications that something resided between the orbits of Mars and Jupiter was drawn through the observational data of Tycho Brahe (1546–1601), the somewhat eccentric Dutch

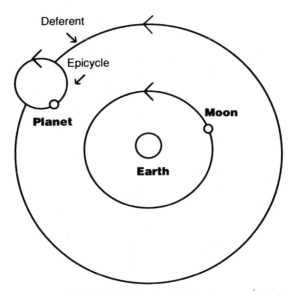

FIGURE 1. A simplified model of the ancient Ptolemaic system of planetary motion: Each planet moves in a small circle called an epicycle; in turn, the planet in the epicycle revolves on the circumference of a circle (called a deferent) around the Earth. According to this system, the Moon and the Sun are the only bodies that do not follow the standard planetary motion.

astronomer who made some of the greatest astronomical observations before the advent of the telescope.

Fate turned Brahe's eyes to astronomy in the form of the brilliant supernova of 1572. Brahe, of noble birth, was arrogant and quarrelsome.[1] In spite of his seemingly nasty disposition, he was able to gather funds from King Frederick II of Denmark and to have an observatory built in Hveen, an island in the Baltic. For 20 years, Brahe made the most accurate measurements of stellar and planetary positions that were ever made prior to telescopes and astrophotography. Brahe would have been the perfect person to find the only visible asteroid, Vesta. There was a chance he did, but he probably dismissed it or mislabeled it as a star.

After the death of his benefactor king, Brahe argued with the Danish court and the new king, Christian IV. Cut off from his funds, he was forced to abandon Hveen, taking many of his instruments and most of his records with him. From there, he went to Prague to pore over his collected data, and became the Imperial Mathematician to the Holy Roman Emperor. A year before his death, Brahe hired spindly German astronomer Johann Kepler (1571–1630) as his assistant to help keep his records of the heavens, and the first very small steps in the search for asteroids began.

In his early career, Kepler had aspired to theology, but his beliefs in mysticism were frowned upon by the church. All was not lost, for he also excelled in the fields of mathematics and astronomy. Kepler held numerous posts as a mathematician, shuffling around from place to place because of political and religious conflicts with others around him. He eventually found his way to Brahe, after sending the famous astronomer (and Galileo Galilei) his *Mysterium cosmographicum*, a geometrical compendium on why God had chosen only six planets (the only ones known at that time—Mercury, Venus, Earth, Mars, Jupiter, and Saturn) to grace the solar system. After all, Kepler believed that God was a geometer who had geometrical reasons for each of the major planets in the system.

Brahe was impressed with Kepler's mathematical prowess and invited the German to Prague as his assistant. Kepler's appointment might strengthen one's belief in fate: Brahe gave Kepler the task of determining a definitive orbit for Mars, for the orbit of the red planet did not fit the geocentric orbital view of the heavens. Moreover, after Brahe's death, Kepler was made the executor of the most accurate astronomical observations collected (mostly from Hveen) up to that time.

Both Kepler's Mars studies and the meticulous measurements of motions of the planets' orbits made by Brahe helped Kepler dismiss the popular idea of circular planetary orbits. In its stead, he formulated the principal laws governing the motion and orbits of planetary bodies based on a closed curve known as an ellipse, thus eliminating the epicycle theories that had governed astronomy for close to 2000 years. Kepler would eventually develop two additional laws concerning the planets: His second law stated that a straight line joining a planet and the Sun would sweep an equal area in space in equal intervals of time. This means that a planet moves faster in its orbit nearer the Sun (or when the planet is said to be at perihelion), and moves slower further away from the Sun (or when the planet is said to be at aphelion). A third law stated that the squares of the sidereal periods of the planets (measured as time) are proportional to the cubes of their orbits' semimajor axes (or their average distance from the Sun).

In fact, Kepler's revolutionary ideas showed that the Earth and other planets moved around the Sun. Perhaps it was best that such a discovery occurred after his mentor Brahe's death. Such a theory would never have been tolerated by the temperamental Dane, who preferred to believe that the Sun moved around the Earth.

Kepler was also one of the first astronomers to suggest that another planet lay between the orbits of Mars and Jupiter, two of the six then-known planets. Using the information from Brahe's observations, he knew that the planets' spacing becomes greater and greater the farther they are from the Sun, with a large gap between the red planet, Mars, and the gas giant, Jupiter. In 1596, Kepler wrote, *"Inter Jovem et Martem interposui planetam"*—"between Jupiter and Mars, I would put a planet."[2]

BETTER FOCUS

One of the greatest leaps in astronomical science took place just over a decade after Kepler's pronouncement: the telescope. The real credit for the invention of the actual instrument has been passed to a spectacle maker in Flanders in 1608, Hans Lippershey (c. 1570–c. 1619). The principle behind the first telescope may have been known to Roger Bacon in the 13th century or to early spectacle makers in Italy. The telescopes were called refractors, with two lenses of opposite curvature and at a certain focal length (or the distance between the center of the lens and the point where all the parallel light rays from the source are bent to a common point or focus). And although Galileo Galilei (1564–1642) is often given credit as the first to use the telescope for astronomical purposes in 1610, it was probably the English mathematician, astronomer, and

physicist, Thomas Harriot (1560–1621), who made the first telescopic observations of the Moon several months before Galileo
turned his telescope to the sky.

No matter who was responsible for its invention or its first
astronomical use, the telescope spurred a race to discover the
details of the sky, especially the immediate neighbors of the solar
system. Christiaan Huygens (1629–1695) was one of the first to
build long refracting telescopes—the longer focal length allowed
for higher magnification. These lengthy tubes allowed Huygens to
discover Saturn's largest moon, Titan, and to realize that the
"ears" reported by Galileo on each side of the planet were actually
rings.

Additional telescopic changes continued to aid in the planetary searches. Johannes Hevelius (1611–1687; Hevelius is the Latin-
ized form of his name; it is also written as Hewelcke and Hevel), a
brewer from Danzig, now Gdansk, was an astronomer and instrument maker. He set up what was then the finest observatory in
Europe and was known not only for publishing the first reasonably accurate drawing of the Moon (1645), but also for his catalogs
of stars and comets.

Hevelius continued to build telescopes with greater focal
lengths, which meant longer telescopic tubes to hold the lenses
farther apart to increase the magnitude. The visual strength of
such telescopes was great, but the physical characteristics made
Hevelius' telescopes become so unmanageable—one of the later
models reached about 45.7 meters in length and was destroyed in
the great Danzig fire of 1679—as to make them worthless as
practical viewing instruments.

Hevelius' huge tubes led Huygens to make tubeless "aerial
telescopes," lining up glass at the top of a tall mast with a hand-
held eyepiece below. Huygens would eventually construct a telescope that reached a focal length of 64 meters. A planetary discovery eluded him, but he did succeed in drawing the first detailed
illustrations of the Martian surface thanks to the longer visual
reach of his telescope.

Giovanni Domenico Cassini (1625–1712), the French astronomer who was the first to determine the rotations of Mars and

Jupiter, was also in the planetary race. He was the first to discover the divisions between Saturn's rings (known as the Cassini divisions) and four small moons of Saturn, which he named the "Louisian stars" (even though they are moons) in honor of his patron, Louis XIV of France. The aerial telescopes paid off at this time, too, as the two faintest moons of Saturn, since renamed Dione and Tethys, were spotted by using such scopes of 30.5 and 41.5 meters.

The discoveries with the newly developed telescopes showed that the solar system was much more populated than anticipated. But in all the searching and discoveries, none of the small "planets" we now call asteroids were ever reported.

CERES BY CHANCE

As the lines of planetary orbits were drawn, more interest in the mechanics of the heavens emerged. The telescopes improved with the advent of reflector (using a combination of lenses and mirrors) telescopes. Although refracting telescopes (using lenses only) were, and are, known for their excellent resolution, higher performance meant larger lenses and tubes. Reflecting telescopes solved these problems by having smaller focal lengths and thus easier-to-manage tubes. In addition, the reflectors did not need such precision within the optics (except one surface had to be precisely ground down), and they were much less expensive to produce.

But it was perhaps an amazing mathematical "law," known as the Titius–Bode Law, that pushed the search for a planetary body that would reside between the orbits of Mars and Jupiter. Until a few decades ago, the law was referred to as Bode's Law, after German astronomer Johanne Elert Bode (1747–1826). But it was actually German astronomer Johanne Daniel Titius (1729–1796) who, in 1766, proposed the idea that the planets seem to fall into orderly orbits. Titius' reference was written in a footnote to a book he was translating, *Contemplation de la Nature* by Charles Bonnet; Bode recognized the significance of the mathematical sequence and popularized the idea in 1772, thus, the initial reference to the law as Bode's Law.

The Titius–Bode Law mathematically explains the mean distances of the planets from the Sun in astronomical units. The empirical formula is straightforward:

$$a = (n + 4)/10$$

in which a is the calculated mean distance expressed in astronomical units of the planet from the Sun and n is the progression of numbers following the sequence 0, 3, 6, 12, 24, 48, 96, 192, and 384. The law falls apart at one point at the edge of the solar system, unable to take in the orbit of Neptune, but comes back to predict the relative position of the planet Pluto. Even today, the law is yet to be explained by any physical argument. And scientists still argue as to whether nature is really that organized.

The Titius–Bode Law in Table 1 produced an interesting observation of the geometric progressions of the known planets at that time—from Mercury to Saturn. But the listing also led astronomers to wonder about the numbers that indicated possible, but yet undiscovered, planets. This made some astronomers skeptical. After all, no one had found another planet past Saturn and there was a gap between Mars and Jupiter. Did the law break down in these two regions?

Table 1. The Titius–Bode Law

Titius–Bode calculation	Planet	Planet's actual mean distance in astronomical units from the Sun
(0 + 4)/10 = 0.4	Mercury	0.387
(3 + 4)/10 = 0.7	Venus	0.723
(6 + 4)/10 = 1.0	Earth	1.000
(12 + 4)/10 = 1.6	Mars	1.524
(24 + 4)/10 = 2.8	Asteroids	2.4–3.0
(48 + 4)/10 = 5.2	Jupiter	5.203
(96 + 4)/10 = 10.0	Saturn	9.539
(192 + 4)/10 = 19.6	Uranus	19.191
—	Neptune	30.071
(384 + 4)/10 = 38.8	Pluto	39.158

William (or Wilhelm) Herschel (1738–1822) broke the stalemate in the outer solar system, discovering the planet Uranus while making a telescopic survey of the heavens. Herschel's finding boosted the belief in the Titius–Bode system: Astronomers measured the mean distance of the planet to be 19.2 astronomical units; checking the Titius–Bode chart, a planet was predicted to reside at 19.6 astronomical units, and Uranus fell well within the predicted distance.

Thus, many astronomers who were skeptical of the geometric progression seized on the idea after about 9 years of virtually ignoring it, and suddenly, another step toward discovering the asteroids was taken. Astronomers were taking a closer look at the solar system and noticed the numerical gap between the planets Mars and Jupiter on the Titius–Bode chart.

In particular, Hungarian astronomer Baron Franz Xavier von Zach, court astronomer to Duke Ernst of Saxe-Gotha and director of the Seeberg Observatory near Gotha, attempted to determine the probable orbit of a missing planet between Mars and Jupiter. Von Zach was tenacious in his systematic sweep of the ecliptic (the plane of the solar system along which most of the planets are located) as he conducted a "revision of the stars in the zodiac," with the ulterior motive of finding the missing planet. By 1788, von Zach held a conference, at which time French astronomer Joseph Jerome Le François de Lalande (1732–1807) suggested dividing the zodiac into sections in the search for the planet. But the immensity of scouring such a wide berth of sky with telescopes was seemingly overwhelming, and Lalande suggested that they turn to the astronomical community to support the effort.[3]

It took two more years before the campaign for a concentrated, cooperative search for a planet between Mars and Jupiter caught on among the astronomical community. On September 17, 1800, even though war in Europe made travel difficult, six astronomers met at the private observatory of German astronomer Johanne Hieronymus Schröter (1745–1816; famous for his lunar observations) in Lilienthal, Germany, and decided to look for a "lost planet." The astronomers—Schröter, von Zach, Karl Harding, Heinrich Wilhelm Matthäus Olbers, Ferdinand Adolf von Ende, and Johann Gildemeister—reasoned that if there was a planet such a relatively short distance from the Earth, it must indeed be dim. Again, the astronomers agreed that the best way to find such an object was a massive searching sweep of the sky.

The Vereinigte Astronomische Gesellschaft (United Astronomical Society) was formed to undertake a systematic search for the planet. The "Celestial Police," as the group nicknamed itself, began with Schröter as president and von Zach as secretary. They divided the sky into 24 zones to accommodate the 24 astronomers of the Celestial Police—including Johanne Bode, Charles Messier, William Herschel, and Johann Huth—seeking objects down to the 9th magnitude. In this case, the term *magnitude* is *apparent magnitude*, or the brightness of an object as seen by an observer without regard to distance, as opposed to *absolute magnitude*, the brightness of an object or a star if it were 10 parsecs from the observer. (To compare, Barnard's star, one of the closest stars to the Earth, has an approximate apparent magnitude of 9.)

The astronomers began their searches, scanning for a planet along the ecliptic at a distance of 2.8 astronomical units from the Sun. In other words, their mission was to try to find a planet in the predicted orbit of the Titius–Bode Law. Because of communication difficulties, it took time for von Zach to contact all the desired astronomers, including astronomer Giuseppe Piazzi (1746–1826).

About 1764, Piazzi was an Italian monk, a member of the Theatine order. By 1780, he became professor of mathematics at the University of Palermo, Sicily, and directed the newly established Palermo and Naples observatories. By this time, too, his reputation as an astronomer was well established, especially in cataloging stellar objects.

Piazzi was not a member of the Celestial Police—at least not yet (although he probably had a general idea that there was a celestial sweep taking place; he joined later). He was conducting his own survey of the sky, telescopically mapping the extra slow motion of stars and measuring stellar positions with extreme accuracy.

One night, Piazzi noticed that a starlike object he had recorded near the constellation of Taurus seemed to have moved; it would move 4 arc minutes right ascension and 3.5 minutes in declination across the sky each night. He also noted that the "star" was retrograding, or moving in an apparent westerly, or backward, motion as seen from Earth in reference to the fixed stars behind it (the usual motion of a planetary body is in an easterly direction). The "star" continued its backward motion until January 12, when it became stationary. Its movements, including its retrograde motion and quick run across the sky, were good indications that the body was located within the solar system.

The initial sighting of the dim object had occurred, strangely enough, on the night of the first day of a new year, January 1, 1801. It was made at Palermo, about 200 years after the invention of the telescope. Piazzi had found the 8th magnitude object, the largest and first-known asteroid, by accident.

KEEPING IT SECRET

Believing he had discovered a faint comet (similar to the mistake William Herschel made upon his discovery of Uranus), Piazzi watched the object (so far, not named an asteroid) from January 1 to February 12, when a "dangerous illness compelled him to discontinue his observations."

By this time, Piazzi had finally sent the news of the newly discovered object to Bode, but referred to the object as a comet. According to O.M. Mitchel, the director of the Dudley Observatory in Albany, New York, in 1868:

> Piazzi not considering it possible that a planet which remained hidden from mortal vision from its creation could be discovered with so little effort as had thus far been put forth, conceived that the moving body which he had discovered was a comet, but the intelligence having been communicated to the society [Celestial Police], Bode promptly pronounced this to be the long sought planet, an opinion in which he was sustained by Olbers and Buckhardt, Baron de [sic] Zach, and Gauss, and I know not by how many other members of the society ...[4]

Although Piazzi made about twenty-two observations of the object, he did not release the details of the sightings until that

August. The reason for his silence was rivalry: He knew the competition to find the planet was keen and he wanted to be sure to secure his right as the discoverer of the new planet. By keeping the planet's positional data to himself, he would have a head start on claiming the planet.

But by the time he released the data to others, the object had gone behind the Sun, into solar conjunction. The Sun's brightness frustrated the Celestial Police's efforts to observe the interloper, and, adding to their frustration, there was no calculated orbit for the object, making the visual recovery of the potential planet highly doubtful.

As so often happens in science, there was one person who was at the right place at the right time. In this case, it was 24-year-old German mathematician Carl Friedrich Gauss (1777–1855), who had devised a technique for calculating orbits using merely three observations. Based on Piazzi's records of the object that were by then available, Gauss calculated where the object would be found as it made its way out of the Sun's rays.

Even with the orbit determined, it took close to a year to find the fledgling minor planet again. Many claim that German astronomers Heinrich W. M. Olbers (1758–1840), one of the original astronomer in the Celestial Police, found the object again. Olbers may have been the first to announce the rediscovery, but it was actually von Zach who recorded the object on December 7, 1801, based on Gauss' calculated orbit. Olbers confirmed the recovery of the object by von Zach—by a few hours. But because Olbers had few stellar maps of the region, he doubted its existence.[5] By December 18, he was almost positive that one of the "stars" had moved; it was not until January 1, 1802 that he convinced himself that it was indeed Piazzi's missing object—one year after its first sighting by Piazzi.

Piazzi named the "major planet" Ceres Ferdinandea, after the Roman goddess of grain and harvests, Sicily's patron goddess, and its Bourbon prince, Ferdinand IV; it was later shortened to Ceres and downgraded to a minor planet. William Herschel was the first to attempt to measure the object's diameter—which he found to be only 260 kilometers (the actual diameter is about 1000 kilometers). Ceres was found within a half a degree of its predicted location and it was a mere 2.77 astronomical units from the Sun—coming close to the prediction of 2.8 astronomical units made by the Titius–Bode Law.

MR. PALISA'S EYESIGHT

Not everyone was satisfied with just the one small planet 1 Ceres (the number accompanying the asteroid's name represents where the object stands in the discovery list; for example, 1 Ceres indicates that Ceres was the first asteroid ever found and confirmed). German astronomer Olbers found the next minor planet in March, 1802, in the same region of the sky. Named 2 Pallas (in the 1800s, this minor planet was also called Olbers) and located at about 2.8 astronomical units from the Sun, it was one of the first definite clues that the region between Mars and Jupiter held more than just one minor planet. Olbers was one of the first to ask if Ceres and Pallas always traveled in their current orbits in the same proximity or were they part of the debris of a former and larger planet—a planet that exploded in a major catastrophe.[6]

Other findings followed, including 3 Juno, by Karl Ludwig Harding in 1804, and 4 Vesta, by Olbers in 1807. By this time, astronomers knew the group of small rocks in the sky were not like the larger planets. William Herschel, with the largest of his telescopes, still could not make out any definition of Ceres' disk, and proposed the word *asteroid* (derived from a Greek expression meaning "starlike"), which was the first time the label was ever used. Other astronomers called the objects "minor planets"—a reference to the bodies' sizes in comparison to the other planets in the solar system. Both terms are still in use today and are synonymous.

Scanning the sky for such dim objects was frustrating, and the Celestial Police disbanded in 1815. Only four asteroids had been discovered, because available telescopes were not powerful enough to catch the reflected light of the asteroids. But more important in the quest for asteroids was the lack of good stellar maps, many times making the positions of possible asteroids a wild guess on the part of the observer.

It took close to 40 years before the next asteroid was found, mainly because of the better stellar maps released in the 1840s. The maps enabled astronomers to find not only more asteroids, but a planet, too. The independent mathematical calculations by British astronomer John Couch Adams (1819–1892) and French astronomer Urbain Jean Joseph Leverrier (or Le Verrier; 1811–1877), as well as the better stellar maps, helped German astronomers Johann G. Galle (1812–1910) and Heinrich L. D'Arrest (1822–1875) at the Berlin Observatory quickly discover Neptune in 1846.

The more accurate maps also boosted the number of asteroid discoveries. By 1845, K. L. Henke found asteroid 5 Astrea and, a few years later, 6 Hebe (1847). John Russel Hind found 7 Iris and 8 Flora in 1847 (he would find 10 in all); 9 Metis was found in 1848 by Graham; and 10 Hygeia was found by De Gasparis in 1849. Hermann Goldschmidt would find 44 Nysa in 1857 (he would find 14 more, some from his window above the Paris Café Procope), and Carl Luther would find 95 Arethusa in 1867 (he would discover a total of 24).

More than anyone else in asteroidal search history, Johann Palisa, who started his career in Pola, Austria, and later worked at the Vienna Observatory, takes the prize: Palisa discovered 121 asteroids, using only star charts and a telescope. No one has made a visual discovery, meaning only with the eye and a standard telescope, of a new asteroid since his death in 1925. Palisa's sheer determination has to be admired. Asteroidal searches were exhausting and inefficient. If a star was not found on a chart, it became suspect as an asteroid. And the only way to prove the existence of an asteroid was to detect its motion, which sometimes took months to verify.

In all, the total number of known asteroids reached 100 by 1868, 200 by 1879, and 300 by 1890. As the discovery of asteroids began to slow down, a new technique helped to boost the numbers once again: Photography was introduced into astronomy.

PHOTOGRAPHY'S BOOMING BUSINESS

Astronomical photography blossomed in the late 1800s. The new medium increased the number of asteroids found. Asteroids also became somewhat of a nuisance, as their trails cut across time exposures usually reserved for long-exposure, deep sky photos of very distant astronomical objects, such as galaxies and nebulas. Soon, because of their apparent abundance, the asteroids became known—not only to deep-sky observers, but the general public as well—as "vermin of the sky."

German astronomer Maximillian Franz Joseph Cornelius Wolf (1863–1932) was the leader in the field of asteroidal photos. Born in Heidelberg, he discovered his first periodic comet in 1884 (a comet whose orbit is well known enough to predict its return), and then turned to seeking asteroids using photography. His first photo find was asteroid Brucia, in 1891, the 323rd asteroid known.

Wolf's greatest contribution was his method of photographing asteroids. The idea was relatively simple. Telescopes with motor drives moved the scope to keep up with the spinning of the Earth; thus, any photograph taken of the night sky with such a telescope would show the stars as pinpoints of light. An object, such as an asteroid, is closer to an observer on Earth than a star. Thus the asteroid would move quicker than the stars and would leave a short trail on the photograph—the trail length dependent on the exposure time and distance of the asteroid. By checking the exposures with a magnifying glass (an asteroid trail on a photographic plate is often only millimeters long), Wolf, and soon many other observers, found traces of more asteroids (Figure 2).

Wolf also conceived the idea of the blink microscope, a device in which two different photographs of the same part of the sky are set side-by-side. To compare the images, the microscope automatically shifts the observer's vision back and forth between the same corresponding parts of the two photographs. If the star patterns are the same on both photos, the observer sees a constant image; if one object has moved slightly, the image of that object seems to jump back and forth as the view is switched from one image to the other. (The device is also used to detect varying brightness in stars and, probably most important, was used by American astronomer Clyde Tombaugh to find the planet Pluto.)

By 1900, other astronomers using Wolf's photographic methods would bring the asteroid total to 452; by 1923, there were more than a 1000 known minor planets. Wolf's catch was about 232 confirmed asteroids (although some claim he found closer to 500

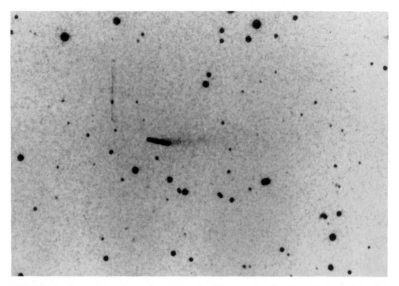

FIGURE 2. One of the "vermin of the sky," a near-Earth asteroid, is seen in this negative. After the advent of photography and its use in astrophotography, astronomers learned that asteroids were more prolific than previously believed. The proof was in the hundreds of asteroid trails bursting across an image, usually on deep-sky photographs. (Photo courtesy of Lowell Observatory and E. L. G. Bowell)

asteroids). But he was far from the top asteroid discoverer who used photos: German astronomer Karl Reinmuth (Wolf's assistant) discovered 385 asteroids between 1914 and 1957; Russian astronomer Nikolai Cernykh found 377; and, more recently, astronomer Edward L. G. Bowell at the Lowell Observatory has counted 386 to his credit, making him the leading discoverer of asteroids that have been numbered.[7]

Asteroids continued to be discovered throughout the 20th century, with at least one asteroid found by various means (except visually, as Palisa did) every year since 1847, with the exception of 1945, the end of World War II. Today, the list of 6000 confirmed asteroids continues to grow. No one could have predicted, especially the astronomers who started the Celestial Police, that, in the voids of the solar system, there would be so many minor planets— so many "vermin" in the sky.

C h a p t e r   2

TIGHTEN THE BELT

*There are no fixtures in nature. The universe is
fluid and volatile. Permanence is but a word of
degrees.*
RALPH WALDO EMERSON
TRYING TIMES

T rying to determine the age of the solar system is not easy.
On Earth, we interpret fossils within bedrock layers, determining Earth ages based on the fossils' positions within
the formation and/or with special rock-dating techniques.
But in the case of the solar system, we do not have such luxuries.
We view the "fossil" surfaces of the planets, satellites, and sundry
objects from afar, forced to interpret ages based on inference and
interpretation of two-dimensional visual images. And if we are
lucky enough, we interpret three-dimensional images based on
the mixing of visual and nonvisual remotely sensed data.

In order to put the beginnings of our own system into context,
it is only fair that we step backward many billions of years to the
origins of the universe, when the gases, dust, and debris that
eventually formed the asteroids had their beginnings. Astronomers infer that conditions before the birth of the universe were
unimaginably chaotic and superdense, with temperatures greater
than 1500 billion degrees Kelvin. The most discussed (although
not always accepted) idea of our universe's humble beginnings
starts with a infinitely compact and singular state, enclosing a
space even smaller than an atomic particle. The compact ball grew,
not in a violent explosion but as a rapid expansion that is often
referred to erroneously (it did not explode) as the "Big Bang."

The evidence for such a theory is based on discoveries in

astronomy and subatomic physics and especially from relativity and quantum theory. With these theories, astronomers are trying to understand the reality of not only the Big Bang, but also what happened fractions of seconds immediately after the Big Bang. One such account includes the following:

- Planck time occurred first, before 10^{-43} second.
- At 10^{-43} second, gravity separated from the three other forces of electromagnetism, strong nuclear, and weak interaction.
- At 10^{-33} second, the three forces were operating.
- At 10^{-10} second, weak interaction and electromagnetic forces separated.
- At 10 microseconds, quarks combined to form particles.
- At 3 minutes, the nuclei of light atoms formed.
- In the first 30 minutes, all the matter in the universe consisted of about 75 percent hydrogen and 25 percent helium.
- At 300,000 to 500,000 years later (the time is debated, as to be expected), the first true, complex atoms formed.
- And finally, several hundred million to a billion years later, gaseous clouds of hydrogen and helium began to condense into protogalaxies and stars (which formed first has recently been debated).[1]

What is the best evidence for the Big Bang? First, galaxies are apparently moving away from each other, indicating that the universe is expanding—what is to be expected if such an "explosion" of energy, space, time, and matter occurred. Second, the radiation reaches us from every direction and is apparently of equal intensity no matter which way we view the universe from Earth.

The third piece of evidence involves chemistry: Scientists believe that helium and deuterium would evolve into a certain specific ratio after the Big Bang—and apparently, the universe holds such a ratio. More recent and better space- and Earth-based telescopes helped determine the ratio: The "fingerprints" for ionized helium were recently found by Arthur F. Davidsen and his colleagues at Johns Hopkins University. The team used the Hopkins Ultraviolet Telescope, part of the Astro 2 Observatory that flew aboard the space shuttle in early 1995, as it monitored the gases that fill the space between deep space (and thus younger) galaxies.[2] Deuterium is a rare form of hydrogen with an extra neutron; it is also called heavy hydrogen and was created naturally only once, during the hot birth of the universe (no stellar cores are hot enough to fuse this material). The 10-meter Keck Telescope atop Hawaii's Mauna Kea was also used recently to measure the abundance of deuterium outside our galaxy, possible now because of the light-gathering power of this, the largest single-mirror telescope in the world. And so far, all measurements seem to indicate the "right" amount of deuterium to support the Big Bang, though more observations are needed to confirm the findings.[3]

Cosmologists also believe that cosmic microwave background radiation is compelling evidence for the Big Bang. Cosmic microwave radiation—or very high-frequency radio waves with wavelengths between 1 millimeter to 1 meter on the electromagnetic spectrum—was discovered coming from space in 1964. Arno Penzias and Robert Wilson, two scientists at Bell Laboratories, were measuring microwave radiation, converting a large radio antenna for astronomical purposes. A background, staticlike hiss appeared to be coming from all parts of the sky. What Penzias and Wilson heard in their receivers was not artificial background noise—or even pigeon droppings on the dish, as they speculated. The ever-present noise, coming from all directions, turned out to be the "echo" (more like a whisper now) of the Big Bang—a reflection of the radiation released from the initial expansion of the universe. (Penzias and Wilson were awarded the Nobel prize in 1978 for their discovery.)

Not everyone assumes that the Big Bang occurred. Probably more than any other model, the steady state model is the most contrary, mainly because it radically counters the idea of the Big Bang. The steady state model, first proposed in 1948 by British astrophysicists Thomas Gold, Herman Bondi, and Fred Hoyle (who, ironically, was the first to coin the term *Big Bang* as a slight against the BB theory), begins with the premise that the universe has no beginning or end. As the universe expands, the distribution of galaxies and other objects (that is, the density of the universe) remains constant, with old galaxies being replaced by new galaxies, making the universe appear to remain unchanged.[4] But the steady state model creates as many pitfalls as it explains. One in particular is the development of new hydrogen protons. How are these protons created spontaneously to keep the density of matter constant?

BEGINNING TO END

Because no being that we know of was here at its inception, cosmologists can only speculate as to the beginnings of our current universe. And as to its fate, again, we can only guess.

It would be nice if scientists could confirm that the fate of the universe rests in the hands, or more accurately, on the back, of a tortoise or some other animal, as some cultures once believed. It would make science so much simpler. But in reality, scientists can lean only on models of the universe to determine its fate, with each model based on more recent (and often more accurate) measurements and observations taken with some of today's best Earth- and space-based telescopes and scientific spacecraft.

Alas, even with all the measurements and observations, astronomers have not developed any real new models of the universe's fate in the past several decades. There have been modifications, but few new models. Many of the theories overlap one another, but there is one thing everyone agrees upon—there is no one theory that is agreed upon by all astronomers. In general:

- In the open universe model, the universe expands toward infinity, and the galaxies, stars, and planets spread out uniformly in all directions. In this model, the universe continues to expand forever, and density throughout the universe becomes less over time.
- In the closed universe model, the universe expands, but gravity takes over at some point, causing the universe to collapse.
- In a "balanced universe" model, the universe just eventually stops expanding.

- In the oscillating universe model, the universe collapses and then rebounds rhythmically over billions of years. Because this model assumes that there is a great deal of matter in the universe, gravity eventually pulls the matter back toward the center; the matter contracts to the point of origin and expands as another Big Bang starts the cycle again.
- In the inflationary universe model, the universe will keep expanding or stop, all based on the presence of dark matter, the theoretical matter that some scientists speculate makes up 90 percent of the universe. If scientists find that there is a low percent of dark matter in the universe, then the universe will expand forever; if there is a high percent of dark matter present, then expansion will eventually stop. So far, scientists have been trying to detect dark matter, often in the form of massive compact halo objects (MACHOs), thought to be giant, dark planets or small burned-out stars that cannot sustain a nuclear reaction. Although there are tantalizing events that seem to indicate the dark matter, there is still no conclusive proof.
- And of course, the steady state theory, where the universe will always be the same density and distribution. According to this theory, the universe 10 billion years in the past and 10 billion years in the future are roughly the same; there are changes, but everything evens out in the end.

The growing evidence from larger Earth-based telescopes, and the orbiting Hubble Space Telescope leans in the direction of the Big Bang, although the model for the fate of the universe is still highly debated, and will continue to be for years to come.

ALL IN GOOD TIME

Trying to determine and agree on the age of the universe is just as tricky as deciding the fate of the universe. The ratio of a galaxy's recessional speed (or how fast it is moving away from us) to its distance is called the Hubble parameter. Also called the Hubble constant, H, and named after American astronomer Edwin Hubble, H is expressed in kilometers per second per megaparsec (about 3.25 million light-years). If the resulting number is high, the universe was born only a short time ago; if the number is low, the universe is much more ancient. Hubble came up with the idea in the 1920s (and others since then have verified his theory), determining that the universe is about 15 to 18 billion years old, dating it from when the alleged Big Bang occurred.

Since Hubble's initial work, refined instruments have allowed cosmologists to make more precise estimates of the actual Hubble parameter. In the early 1970s, Allan Sandage of Mount Wilson Observatory and Swiss scientist Gustav Tammann estimated the Hubble parameter. Their estimation was based on gathered data from well-known techniques and methods (for example, the Cepheid variable methods), and they threw in additional data to augment their research, such as the relationship between the absolute luminosity and a galaxy type, among others. Using this information, they determined the distance to the Virgo cluster, the largest large group of galaxies known, thus approximating the Hubble parameter.

By 1977, Brent Tully at the University of Hawaii and Richard Fisher at the National Radio Astronomy Observatory found a way to determine the Hubble parameter with more precision, by discovering that the width of the 21-centimeter line in the neutral hydrogen spectrum (it helps determine how fast a galaxy is rotating) had a direct correspondence to the inherent brightness of the galaxy. In other words, the brighter the galaxy, the faster the spin and the broader spectral lines appear at particular wavelengths—the Tully–Fisher relation.

Sandage and Tammann then used a modified version of the Tully–Fisher relation and looked again to the Virgo cluster of galaxies. The result was that the universe was close to 18 billion years old (about 50 kilometers per second per megaparsec with a 10 percent uncertainty), a number that was usually accepted by the astronomical community. It fit with many of the cosmological models of the time: The low critical density pointed to a closed, not open, universe. The result also fell comfortably into the idea that our Sun (and thus the solar system) is a baby compared to the deep-space stars that occupy the outer regions of the universe.[4]

But in actuality, the Hubble parameter is an elusive beast of a number, primarily, according to many scientists, because the measurements have such large systematic errors. And more recent work has thrown in another cosmic monkey wrench: In 1978, more work was done on determining the Hubble parameter by Gerard de Vaucouleurs at the University of Texas, Austin. He determined that the Hubble parameter was about 100 kilometers per second per megaparsec (with about 10 percent uncertainty). If this is correct, the universe is only 9 billion years old and gives points to those astronomers who believe in the open universe models. Other astronomers have also added their numbers to the foray, ranging from 40 to 90 kilometers per second per megaparsec, all definitely turning toward a younger universe.

The story continues to this day, with data from the Hubble Space Telescope gathered in the mid-1990s further perplexing scientists. Galactic cluster distance data from the Hubble were recently analyzed by a team of astronomers headed by Nial R. Tanvir at the University of Cambridge in England, who used the data to calculate a Hubble parameter between 61 and 77 kilometers per second per megaparsec, or about 9.5 billion years.[5] If this is true—or if any of the other recent estimates of the universe's age, ranging from about 8.4 to 10.6 billion years old, are—there is definitely a scientific conundrum. The universe would be even younger than its reportedly oldest stars, thought to be between 13 to 16 billion years old. Logically, this causes consternation among those in the scientific community—how can the universe be younger than its oldest stars?

Is the universe really as young as the Hubble Space Telescope data and other astronomers' calculations indicate? Will cosmologists have to change the way they view the models of the universe—or change the models completely? Are the estimates for the ages of stars not as accurate as once thought? How reliable is the Hubble constant or the methods of determining the constant? And will scientists find out that there are different expansion rates closer to home than in the more distant parts of the universe?

If it is determined that the universe is younger, cosmologists may have to revise all ages in the universe, including the age of our Milky Way. It would be truly phenomenal if the universe were only 8 billion years old, meaning that our about-6-billion-year-old solar system evolved very quickly after the Big Bang. And if so, we may have to further revise our ideas on how and why our solar system and its members formed.

BACK TO THE SOLAR SYSTEM

If current models of the age of the universe are correct, our Milky Way galaxy is thought to have taken shape about 14 billion years ago. The galaxy started as a collection of material spinning around a central area, with tendrils of dust and gas collecting from the center outward to create crude spiral arms. Wherever clumps of the dust and gas resided, stars started popping up—even before our own Sun made an appearance. Some of these stars are listed in modern stellar atlases; others are now remnants from better days, burned cinders or exploded into particles spread out in the voids of interstellar space.

In most places, the concentration of dust and gas created stars composed mostly of hydrogen and helium, the life-gases of stellar furnaces. Nuclear reactions within the stars created different atoms, including the much heavier atoms such as silicon, carbon, oxygen, and iron. And when one of these stars exploded into a supernova, the atoms reached out into space, with enough motion from the explosion eventually generating more stars and their related solar systems. The continual buildup of these atoms enabled everything around us to develop—every animal, mineral, and flower that did, does, and will exist.

What we now call our solar system is much younger than the Milky Way, and it began forming around 10 billion years ago. Initially, the denser parts of our galaxy's interstellar medium (loose clouds of dust and gas) created a nebula (tighter clumps of dust and gas), which remained stable until something compressed it. Like many other such clouds, it probably fragmented in one of two ways (although there are probably plenty more ways to fragment not yet discovered). It was disturbed either by the natural movement around the galaxy's central disk or by the shock waves emanating from a supernova. The atoms were shoved and pushed, squeezed by the hammering waves, until clumps of the dust and gases in the cloud began to pull together, attracted to each other by the force of gravity. The nebula then fragmented into even denser regions. These nebula fragments formed what is often called cocoons—debris clouds around a newly forming star, or the incubator in which stars are eventually born.

In general, our primitive pre-solar system solar nebula of gas and dust is thought to have formed from such an occurrence, and was about twice (others say 100 times) the size of today's solar system. As it began to contract under the influence of its own gravity, the mass concentrated in the center and created the precursor to our Sun, called the proto-Sun.

Like all bodies in space, the original nebula was slowly rotating; as the cloud's center contracted, it began to rotate faster, conserving its angular momentum. The total angular momentum of a cloud depends on the mass, radius, and rotational velocity of the object. In this case, a change in one variable resulted in a proportional change in the other two—as the nebula collapsed, it conserved angular momentum by rotating faster. This forced the rest of the cloud to resist forming a sphere and created a flat disk rotating along the same plane around the proto-Sun.

The flattened, hot cloud of dust and gas continued to spin, shrinking into a smaller, tighter disk. The central disk became hotter, the proto-Sun radiating as a result of gravitational energy lost during its contraction. And after a few tens of millions of years, the core "ignited" from a hot, nuclear reaction.

Our star was born, or so many scientists believe.

The Sun was not the only body to form. In the outer regions of the spiraling nebula, other processes were taking place. Parts of the thin, flat disk of debris that stayed on the outside of stellar formation did not go to waste: The cocoon lost some of its material to interstellar space as the result of natural attrition. As the solar nebula became less dense over time, it also allowed heat to be radiated away more readily, eventually leading to the condensation and aggregation of material from the nebula.

The heat differential also created a temperature gradient

across the disk. Around the center, the highest temperatures reached about 2000 degrees Celsius; at its very edge, about at the outer rim of our modern solar system, the temperatures reached about minus 266 degrees Celsius—similar to temperatures at today's rim. Such differences in temperature became important during the formation of planetary bodies, creating very different chemical conditions (and thus different planetary compositions) from the inner to the outer solar system.

For example, chemical composition studies give us clues to the reason for the clumping within the original nebula and where certain asteroids and meteorites were formed within the solar system. One way is to analyze specific isotopes, forms of the same elements, but with differing numbers of neutrons (in other words, the atomic number is the same, but the isotopes have a different atomic mass and physical properties). In particular, oxygen-16 and aluminum-26 detected in meteorites point to a possible culprit that started the formation of our solar nebula: a supernova, as both isotopes are produced during such superviolent explosions.

Isotope abundance or scarcity in an object can also tell scientists where in the solar nebula the meteorite and its parent body (usually thought to be an asteroid) may have formed. In general, the inner solar system was homogenous, in which the higher temperatures vaporized the interstellar dust grains, leaving the interstellar atoms thoroughly mixed with the rest of the elements in the disk. In the outer solar system, the temperatures were too low to vaporize the interstellar grains, creating a more heterogeneous environment. Such grains are often found incorporated into calcium-aluminum-rich inclusions (called CAIs) and chondrules found in meteorites that land on the Earth.[6]

Thus, large and small planetary-type bodies began forming in the new solar system. The solar nebula, thought to contain many of the elements seen on the surface of today's Sun, was abundant in hydrogen, helium, oxygen, carbon, nitrogen, neon, silicon, and, to a lesser degree, magnesium, iron, and sulfur. Using the more primitive meteorites as a guide, scientists believe that certain elements condensed: As the cloud cooled, elements with the highest condensation temperatures formed, such as the oxides and silicates of aluminum, calcium, and titanium; others followed, including iron and nickel; and next came the rocks filled with such minerals as magnesium-rich silicates. At even lower temperatures, more volatile elements condensed, such as iron sulfides; and although oxygen contributed to many of the condensates, hydrogen also began to cling to other elements, such as oxygen to form water (H_2O) and nitrogen to form ammonia (NH_3). Carbon was also included in the condensations, but it is less understood which molecules actually formed in the early solar system, such as carbon monoxide (CO) or methane (CH_4) (carbon combinations are more affected by temperatures and the amount of other elements available).

The combinations of these elements eventually evolved into the more familiar planetary bodies rich in either silicate minerals, ices, gases, and carbon compounds, or combinations of them. But it would be a long time before the actual planets would truly mature to resemble modern planets. In the meantime, the condensed materials were becoming unimaginably crowded together.

CROWDS OF THE EARLY SOLAR SYSTEM

The modern solar system is pristine compared to what it was like in its early history about 5 to 6 billion years ago. Scientists believe that the early solar system was much more crowded then with amounts of material difficult to imagine. The sky was filled with early versions of the asteroids—icy and rocky bodies that formed from the solar nebula, ranging in size from dust particles to large chunks hundreds of kilometers in diameter called planetesimals. Some of the masses became progressively larger on their way to becoming the protoplanets, the beginnings of today's planets. This occurred through adhesion, as the random motion of the particles in the condensing cloud caused the particles to collide and adhere together; and later, through the gravitational pull of the larger masses on the smaller pieces, caused the objects to collide and adhere—growing even larger as they attracted the smaller debris of the solar system.

As to how fast or slow this accretion took place in comparison to the cooling nebula, scientists can only speculate. Some believe in the heterogeneous accretion model, in which the cooling rate was slow, allowing the less volatile material to form the proto-planets; others suggest the homogeneous accretion model, in which the cooling rate was much higher, allowing the accretion to take on particles with a whole range of compositions. The reality is probably between the two models, as either theory explains the decreasing densities of the planets outward from the Sun.

Whatever accretion model took place, the result was an excessive amount of material circling the central Sun—and perhaps, as some scientists speculate, the inner solar system was filled with asteroidlike bodies. Like a giant traffic jam in space, such crowding of the small and large bodies would not last. Eventually, the accumulations of rocky debris would be eliminated: Material was swept up by the larger forming planets, leaving their orbits devoid of asteroids. Other small bodies were perturbed by the forming planetary bodies, sending the asteroids into eccentric orbits, and often into a collision path with a larger body; or gravity of the larger forming bodies kicked the objects out of the solar system altogether.

Amidst all the activity, the planets and satellites continued to form. (Table 1 gives the distribution of mass in today's solar system; the total mass of the asteroids is also often estimated as one-tenth the mass of the Earth.) And not one body in the inner solar system was exempt from the invasion of the rocky debris. In fact, based mainly on the evidence from the Moon and Mercury, a massive bombardment from these rocky chunks of debris apparently took place in the inner solar system, collectively called the Late Heavy Bombardment (LHB).

The LHB apparently occurred from about 4.0 to 3.8 billion years ago (although some estimates range from 3.5 to 4.2 billion years). The rate of the bombardment seems to have diminished rapidly about 3.8 billion years ago; and based on ejecta blankets on the Moon (the debris thrown up by the impact), the impact rate in the inner solar system has apparently remained somewhat constant since at least 3.5 billion years ago. Before that time, there were many planetesimals striking the inner and outer planets, the

Table 1. Distribution of Mass in Today's Solar System

Object	Percent of mass
Sun	99.86
Planets	0.135
Satellites	0.00004
Comets	0.00003 (estimated)
Asteroids (minor planets)	0.0000003 (estimated)
Interplanetary medium (dust and gases)	less than 0.0000001 (estimated)

planets gaining in mass quickly (geologically speaking) as they swept up the asteroidlike debris. Some collided with other planetesimals; others slammed into planets and satellites, breaking through weak cracks in the forming crusts or icy surfaces (see figures 1, 2, and 3).

Not all the flotsam and jetsam was taken up by the larger or smaller bodies of the solar system. There were bits and pieces that never danced too close to a moon or planet. The majority of the debris was swept up, leaving little dirt in the solar system dustbin. But in its incomplete methodology, the solar system had leftovers.

HOW DO WE KNOW?

The giant impacts of the LHB were some of the most important events that took place in the early solar system. The understanding of the giant impacts first became apparent in the early 1960s, as astrogeologist Eugene Shoemaker and others worked on maps for the Apollo lunar landing missions. Basing their interpretations on superposition (in which the younger rocks overlie the older rocks) and a relative-size frequency distribution of lunar craters per unit area, the researchers were able to determine the relative ages of various spots on the Moon. Radiometric dating of material from the Apollo mission confirmed their findings: Some of the older regions turned out to be the rugged lunar highlands, remnants of huge impacts that occurred about 4 billion years ago. In addition, the lunar maria—which are younger than the high-

FIGURE 1. This Mariner 10 image of Caloris Basin on Mercury shows evidence of the Late Heavy Bombardment, a time when planetesimals, large asteroidlike bodies, smashed into the planets of the inner solar system. The Mariner craft was only able to image half of the 1300-kilometer basin. (Photo courtesy of NASA)

FIGURE 2. The Late Heavy Bombardment also affected the planet Venus, although the results of the heavy strikes are not as noticeable as on the Moon and Mercury. The main reasons for the fewer impact craters are similar to those relating to the Earth: A thick atmosphere and long history of volcanism and planetary activity wiped out many of the craters. This photo shows Sachs Patera on Venus, a sag-caldera formed by volcanic action; the dearth of impact craters around the site is indicative of a planet with a relatively young surface—a surface that has eliminated many traces of heavy bombardment. (Photo courtesy of NASA)

FIGURE 3. Our own Moon did not escape the Late Heavy Bombardment. The impacting bodies broke open parts of the newly formed crust, which in turn released gigantic lava flows. The bombardment continued, and because there is no atmosphere or crustal movement, the resulting (and subsequent) craters still cover the entire satellite—virtually a fossil remnant of the bombardment. (Photo courtesy of NASA)

lands and may have precipitated the volcanic flooding of the major impact crater basins—developed from a different population of impacting bodies.

Extrapolation of the dates on the Moon also led to closer looks at crater-covered Mercury and more sparsely cratered Mars. On both these worlds, crater counts revealed the same message as on the Moon. A huge bombardment took place before about 3.8 billion years ago. In fact, when analyzing the relative size–frequency distribution of the lunar highlands and Mercury, they are very similar, which means that they were probably cratered by the same population of impacting bodies. Even more recently, a detailed analysis of the ratio of argon-39 and -40 in one of the oldest Martian meteorites found on Earth (meteorite ALH84001) suggested that the rock had gone through extensive bombardment. About 4 billion years ago, say Grenville Turner and his colleagues of the University of Manchester, England, the argon escaped as the rock grew hot from catastrophic impact.[7]

Here was evidence that the LHB was a major player in the early formation of the inner solar system. But the outer solar system is trickier to interpret and may have fared differently during the LHB. At this writing, there is little evidence confirming that the same LHB took place at the same time as in the inner solar system, or that one ever occurred at all. Many astronomers point to the broken and cratered surfaces of the outer solar system moons. But many of those strikes could have occurred over time, as the larger planets have a tendency to drag straying comets and asteroids into their gravitational well toward the parent planet and its satellites.

Thus, asteroidlike impacts were literally one of the most significant processes taking place in the early solar system, and one that eliminated many of the asteroids. But it wasn't only impacts that shaped the number of asteroids; there were also the larger, planetary evolving bodies taking a toll on the smaller objects of the solar system.

THE PLANET THAT NEVER WAS?

O ruined piece of nature! This great world
Shall so wear out to nought.
WILLIAM SHAKESPEARE
KING LEAR

PLANET SMASHING, ANYONE?

Early in the solar system's history, besides the asteroidlike planetesimals in the rest of the solar system, the region between the planets Mars and Jupiter contained an extensive collection of objects. A great ring of small planetoids—larger than today's asteroids, yet smaller than the planets—orbited the region, colliding, shattering, and merging eventually to form the asteroid belt.

When the asteroids were first discovered, scientists raced to theorize about their origins. One of the favorite theories was that the asteroids were leftovers from an exploded planetary body. According to M. O. Mitchel, director of the Dudley Observatory in Albany, New York, in 1868[1]

> In case these speculations were within the limits of the probable, and if it were permitted to anticipate in the future, the possible collision or union of these minute planets, a like train of reasoning, running back into the past, would lead to the conclusion that in case their revolution had been in progress for unnumbered ages, there was a time in the past when these two independent [asteroid] worlds might have occupied the same point in space, and hence the thought that possibly they were fragments of some great planet, which, by the power of some tremendous internal convulsion, had been burst into many separate frag-

ments. This strange hypothesis was first propounded by Dr. Olbers, and has met with more or less favor from succeeding astronomers, even up to the present day, as we shall see hereafter.

Blowing up a planet often is a central theme in science fiction stories—as a science fiction writer, I have written it myself. But in reality, exploding a planet into chunks that neatly form an asteroid belt is rather difficult, primarily because no known internal source of energy is great enough to crack what we consider a planet and cause its fragments to scatter. It is true that pieces of a planet can be whacked off if a huge impactor strikes at the right angle and in the right place—especially in the planet's formative stages. But more often, if a planet were blasted to bits, the most likely candidate being a large planetoid striking the planet in just the right place, the pieces would be pulled back by the planetoid's own gravity and coalesce into a fragmented planet. In fact, Uranus' tiny satellite Miranda may have formed and reformed at least five times in this way. The moon probably broke apart from large impacting bodies, then pulled back together because of the gravitational attraction of the fragments (see figure 1).

The second reason for rejecting the "smash the planet" scenario for the asteroid belt is mass: The current number of asteroids within the belt would not equal enough material to form a planet. According to estimates, a body composed of the asteroids thought to have occupied the main-belt billions, or even millions, of years ago would only be about 1300 kilometers in diameter. The body would hardly be large enough to form what we perceive as a genuine planetary body.

Third, the compositions of the asteroids differ greatly, and we base that information mainly on the spectroscopic analysis of the bodies. In particular, it is thought that a planet would have more homogenous, or similar, material representative of its original composition. But in reality, asteroids are made of different materials depending on their position in the asteroid belt. If the belt originated from an exploded planet, the planetary material would be more scattered and less orderly. Each asteroid appears to have condensed from the original solar nebula, as did the planets, its

FIGURE 1. Miranda is one of the smaller moons around Uranus. It is also one of the strangest bodies in the solar system, apparently torn and tortured by impacts. Many scientists suggest that the fragmented features and ragged ridges are evidence that the moon has broken apart and reformed at least five times, probably as a result of impacts by large bodies that struck in its early history. (Photo courtesy of NASA)

composition dependent on the nebula's temperature in the region where the asteroid formed.

The fourth piece of evidence that the asteroids could not have come from a large planet is strictly comparative, deduced by comparing the spectra of asteroids and those of meteorites found on Earth. Most of these chunks of rock are thought to have originated in the asteroid belt, and scientists have found that no minerals within meteorites have undergone high pressures during their formation. Planetary bodies should exhibit some evidence of higher pressures that occurred during the process of formation. There is a chance that such pressures could have been reached on the asteroids Ceres and Vesta. But it is difficult to verify, because apparently only a few of the meteorites found on Earth originated from these two bodies. (There is one exception to the high pressure scenario: Minidiamonds have been found in some meteorites, but they probably formed from high-speed impacts with other bodies, not from the formation of a parent body.)

In reality, the formation of asteroids at the dawn of the solar system makes these rocky bodies the most sought-after pieces of planetary history in the system, true planetesimals that represent the very beginnings of our system. But why did the asteroids elect not to form a planet?

BLAMING JUPITER

We all know the saying, "the squeaky wheel gets the grease," and in the case of Jupiter, it appears to be the same: The large planet gathered much of the solar system's extra material.

The beginnings of the asteroid's main-belt between Mars and Jupiter was not the result of one single event, but many strung together in sequence, all involving the giant gas planet. The first episode in the asteroid belt story starts with an event in the solar system's early history: The T-Tauri phase of the Sun, a time when the young Sun produced powerful winds that threw material outward from the star. (The so-called "freeze line," the point where volatile materials from the solar nebula were unable to condense, also played a part in the formation of the gas giant, but the T-Tauri was one of the most important phases).

Younger stars are known to have very strong stellar winds. In the case of our star, this stage, as noted, is often called the T-Tauri phase of the Sun. The fierce stellar winds blew out from the Sun in all directions, blowing most of the gases and dust from the inner solar system and into the outer solar system. The T-Tauri phase can be somewhat visualized on the Sun today, to a localized extent in the form of a solar flare, a localized brightening of the Sun's surface that quickly ejects matter into space. It is almost as if the Sun is "blowing" the material outward (the actual flare is caused by the interaction between the star's rotation, convection, and magnetic field). To visualize the early T-Tauri phase of the Sun, imagine much stronger solar flares occurring all over the surface and spreading outward.

The T-Tauri phase is thought to be responsible for blowing away tiny (compared to the other planets in the inner solar system) Mercury's atmosphere; and possibly many of the volatiles, or materials that readily evaporate, such as water, existing on the asteroids within the forming asteroid belt. The large sizes of the forming inner planets—Venus, Earth, and Mars—helped those planets keep their atmospheres to a certain extent. The T-Tauri phase was also partly responsible for Jupiter's evolution. The planet was already one of the largest clumps of material forming from the solar nebula. As streams of material flew from the T-Tauri phase of the Sun in Jupiter's direction, the giant planet gravitationally collected many of the chunks, giving Jupiter the most mass of any of the planets. Thus, Jupiter became the controlling body in the outer solar system, which included keeping the asteroid belt in its place.

The T-Tauri event contributed to the growing Jupiter, which was then responsible for arresting a planet's growth between the orbits of the gas giant and Mars. As the planetesimals were accreting and growing elsewhere in the early solar nebula, several dozen of the larger objects called planetesimals (ranging in size from about 100–1000 kilometers) orbited the Sun near the forming Jupiter. No other region within the solar system had such a formidable planet as Jupiter forming, and thus, no other asteroid belts formed. But around what was to become the gas giant of the solar system, a planet was not "allowed" to form.

What were the conditions that persisted in the early asteroid belt? One theory states that several large planetesimals passed too close to Jupiter. Gravitationally scattered into eccentric orbits, the planetesimals fell into the present asteroid belt region. After several close encounters with the forming asteroids of the belt, the planetesimals were thrown out of the solar system by the gas giant, but not before they created quite a stir in the asteroid belt. The close encounters caused the asteroids to increase their velocities and even alter their orbits so they could not form a planet. Such chaos within the belt increased the number of collisions between themselves and perhaps a few of the planetesimals that caused the problem in the first place.

Another plausible theory again involves massive Jupiter. Instead of coalescing into a planet, the perturbation of Jupiter's gravitational field alone tugged at the bodies forming in the belt, not allowing the objects to form into a planetary body. The pushes and pulls increased the acceleration of the many asteroids, causing a greater chance of collisions. Instead of accumulating into a planet, the forming asteroids began to strike each other at speeds of kilometers per second, breaking each other apart. The collisions also spread meteoroid fragments, many of which eventually reached the Earth and other planets and satellites as meteorites.[2]

Even after the formation of the early solar system, the gravitational attraction of Jupiter's giant mass continued to prevent material from accreting into one large body in the region just inside the planet's orbit. What was left was the rocky asteroid belt, also called the asteroid main-belt.

Over time, the Jupiter-influenced nudges and pulls increased asteroidal velocities within the main-belt, causing violent collisions between the bodies and creating the angular material (and many of the meteorites) we see today. And not all the asteroids stay within the asteroid belt for very long in terms of geologic time. Calculations prove that the gravity of Jupiter redirects many of the objects from the belt, sending the rocks out of the solar system or into the inner solar system, bodies that often wind up as near-Earth asteroids.

Maybe it was a good thing that the asteroid belt left planet forming to the other members of the solar system. After all, if all the asteroids in the main-belt were to coalesce into a planet, it would be about the size of our own Moon or one of Jupiter's larger Galilean moons, and Jupiter would easily drag such a small body into its orbit. Scientists believe that at one time there were more asteroids, but that years of pushing, pulling, and colliding in that part of the solar system eliminated many of the asteroids. It is thought that more than 10,000 asteroids exist today; in the ancient past, about 5 billion years ago, there were probably hundreds of thousands more. Some of the material was ejected into interstellar space by Jupiter's gravitational influence, while other material wandered the solar system in chaotic orbits and eventually collided with larger planetary bodies.

HOLDING THE MAIN-BELT

At a casual glance, today's asteroid main-belt seems so orderly: The majority of the asteroids revolve around the Sun counterclockwise, in the same direction as the principal planets. Similar to most of the planets, the mean inclination of the asteroid orbits is about 9.5 degrees from plane of the ecliptic (the plane that most planets, including the Earth, follow in their orbit around the Sun), and they exist between 2.0 to 3.3 astronomical units from the Sun. Like a tightly packed swarm of bees, the asteroids within the main-belt stay in their "place," at least for the most part.

Obviously, it is not the hive mentality that keeps the asteroids so orderly; the reasons are physics and Jupiter, both of which create lanes thick with asteroidal material and expansive gaps within the main-belt. The gaps were first correctly explained by Daniel Kirkwood in 1866 as being due to the perturbation effects of Jupiter; they were later named for him.

The Kirkwood gaps (and thus, indirectly, the crowded lanes) are caused by a perturbation called *gravitational resonance*. The asteroids are in resonance with Jupiter's gravitational pull or are commensurable. In other words, the gravitational attraction of Jupiter constantly adjusts the orbits of the asteroids into certain configurations. Thus for millions, if not billions, of years, such strong gravitational pulling by Jupiter at these geometrically placed positions prevented asteroids from residing for long in these hazardous zones.

Resonance is when the orbit of one body is commensurate with the orbital period of another body, with the two orbital periods being whole number multiples of each other. In the case of Jupiter and the asteroids, the gas giant "sweeps clean" certain lanes within the asteroid main-belt because of this resonance. For example, when an asteroid enters a gap area, it is eventually forced out: When the asteroid is adjacent to the giant planet, it is gravitationally attracted to the planet and "swings" toward the planet; when it is farther away it "swings" in the other direction. The gravitational attraction at each close planetary pass is additive, and with enough contacts, the asteroid's motion (or its "swinging") becomes chaotic. The swinging asteroid within a Kirkwood gap eventually has a revolution period around Jupiter that was an exact fraction of the gas giant, and eventually, the small object is thrown out of the Kirkwood gap (see figure 2).

The number of major gaps is often debated, although most astronomers list between 7 and 10; there are three especially prominent lanes found at 2.5, 2.83, and 3.28 astronomical units from the Sun. Daniel Kirkwood correctly theorized that the distances corresponded to the orbital periods that are simple fractions of Jupiter's orbital period of 11.86 years. For example, the gap at 2.5 astronomical units represents an orbital period of 3.95 years. This is exactly one-third of Jupiter's orbit. An asteroid at this distance would be under Jupiter's gravitational influence, as it lines up with the gas giant every three asteroid years. The other gaps also have direct ratios with Jupiter's period: For example, at 2.83 astronomical units, the ratio is 5:2 (with a period of 4.76 years); at 3.28 astronomical units, the ratio is at 2:1 (with a period of 5.94 years).[3]

All of the ringed planets, Jupiter, Saturn, Uranus, and Neptune, exhibit resonance with their surrounding neighbors, but the gaps in the rings of Saturn are the best examples of the Kirkwood gaps in miniature (the planet's rings are much smaller than the asteroid main-belt). For example, the Cassini's division, between the two outer rings, is caused by the satellite Mimas. A particle traveling in the Cassini's division would have a period of just half

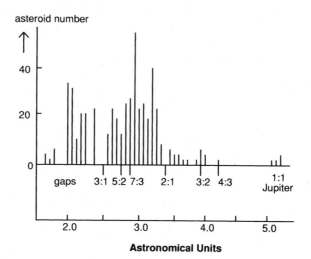

FIGURE 2. The Kirkwood gaps fall into a set pattern throughout the asteroid main-belt. The gaps form in response to the resonance with the gas giant, Jupiter (resonances are shown here as ratios). The more defined, labeled units form what are called "families" in the asteroid main-belt (not to scale).

the period of Mimas, making the particle at the same place in its orbit every second time around. Over time, Mimas would tug on the particle, gradually removing any material from the region, which is similar to the effects Jupiter has on the asteroid main-belt (see figures 3 and 4).

The larger satellites of Saturn may explain the bigger gaps within the ringed system, but it does not explain the thousands of smaller rings, or ringlets, found by images from the Voyager spacecraft taken in the late 1970s. Apparently, there may be other reasons, such as density waves caused by the smaller satellites of the planet, which create the waves by the resonance of the ring particles and a Saturnian moon. There are also "shepherd satellites"— tiny moons seemingly caught within the ring system that nudge straying particles back into the ring system. These nudges thus create the additional, albeit smaller, gaps and strangely kinked lanes of ring material (see figure 5)

Jupiter does not always perturb or throw out asteroid debris

FIGURE 3. The numerous gaps in Saturn's rings developed much the same way as the gaps in the asteroid belt, by falling into resonance. In the case of Saturn's rings, the gaps are created by being in resonance with the surrounding small satellites of the Saturian system. (Photo courtesy of NASA)

in every direction from the main-belt. Occasionally, there are asteroids that escape the belt and remain at a respectable distance from Jupiter, and some reside outside the 2:1 resonance with Jupiter, at about 3.3 astronomical units. With a few exceptions, these asteroids are found in three distinct zones beyond the asteroid main-belt: The Cybeles are between 2:1 and 5:3 Jovian resonances (with

FIGURE 4. Also similar to the asteroid belt gaps, the lanes of Neptune's rings reflect resonance with the planet's satellites. There are also clumps within the rings, thought to be caused by the influence of the moons. (Photo courtesy of NASA)

an average orbit of 3.4 astronomical units); the Hildas are found at 3:2 resonance (with an average orbit of 4.0 astronomical units).

The most interesting grouping in particular are the Trojans, or Jupiter asteroids, the name given to two groups of asteroids that follow the same orbit as Jupiter, following (or leading?) like minions to the gas giant. They are located at the Lagrange points 4 and 5 (or L4 and L5 points). One group orbits 60 degrees ahead of Jupiter, called the Achilles Group; and the other group orbits 60 degrees behind Jupiter, called the Patroclus Group—named after the combatants in the Trojan war. In this case, the asteroids have become locked into an orbit at the same mean distance from the Sun and the same orbital period as Jupiter, in a 1:1 resonance.[4]

The reason for the 60-degree splits can be explained with the theoretical solution of the three-body problem formulated by the French mathematician Joseph Louis Lagrange (1736–1813) in 1772. The points, also called Lagrangian (L) points and referred to as potential minima, are actually maxima in the gravitational potential field (the objects stay, relatively, at the L4 and L5 points). Lagrange proved theoretically that an object that is located equidistant from both the Sun and Jupiter would be stable, traveling around the Sun with the same period as the gas giant. The majority of Trojans follow this rule, with most of them being 20 degrees of the vertex of an equilateral triangle, with Jupiter and the Sun at the other vertices, and all moving about the vertex in a complex curve.

But as in all the bodies of the solar system, there are exceptions to the Trojan 60-degree rule: Some Trojan asteroids can move an appreciable distance away from their mean position, or vary their distances from the Sun. For example, Trojan asteroid 1437 Diomedes can be as far away from Jupiter as 36 or 100 degrees. There are two main reasons for the eccentricities in Trojan location: First are the libration amplitudes that cause the asteroids to fall within 30 degrees of either side of a Lagrangian point; second, a high proportion of these asteroids have high inclination orbits. Both these factors cause the Trojans to be anything but tightly packed in the sky, spanning a wide area in degrees.

Overall, the most important part about the asteroid main-

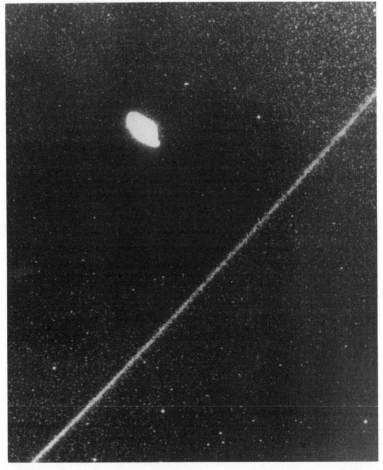

FIGURE 5. Saturn's moon Prometheus is a small "shepherding" satellite; it is about 100 kilometers in length and is probably made mostly of water ice. The moon (and other moons like Prometheus) keeps several of the narrow ringlets of the ring system in line. (Photo courtesy of NASA)

belt's Kirkwood gaps is the fact that these asteroidal alleyways are almost always kept clean. And where does all the material from this cosmic dustbin go? The universe does not discriminate: The elimination of these objects is a perfect way for asteroids and smaller meteoroids to be thrown not only out of the solar system, but also toward the inner solar system to become asteroids and meteoroids that come close to the Earth.

Chapter 4

SHAPE OF THINGS TO COME

Atoms or systems into ruin hurled
And now a bubble burst, and now a world.
ALEXANDER POPE
CATASTROPHIC LIVES

Marie Morisawa was my mentor in high school (although she never realized it), and, during a long-overdue session with graduate school at the State University of New York at Binghamton, she became one of my thesis advisors. Marie was a pioneer in geomorphology, the study of landform development, and a true legend among those of us who worked with her. Most of her work dealt with fluvial geomorphology, the study of landforms resulting from the water that digs into the Earth and creates much of past and present surface geology. But she also was a true overall geomorphologist, helping us to understand the basic concepts that carve a river bank, create a volcano, or hew a submarine canyon.

Marie was not a geocentric geologist, and she encouraged all of us to understand the surface workings of other planets and satellites. We were fortunate enough to live on the Earth, with its winds, waters, and other physical and chemical types of erosion and deposition. She also realized that other planets and satellites had their own characteristic morphological controls, too, some even more dynamic: Violent impacting, extremely active volcanics—even cracks in an icy crust gushing out slush, like squeezing toothpaste from a tube.

During one of our long talks, she once stated a simple axiom that has stayed in my mind all these years: "Morphology is everything."

Morphology, or the measurement and mathematical analysis of the configuration of a space body's surface and the shape and dimensions of its landforms, is everything in interpreting the asteroids, too.

The two asteroids we have so far visited by unmanned spacecraft (Gaspra and Ida), or the moons around other planets that are suspected captured asteroids, do not exhibit Earthlike morphological features. So far, we have not detected meandering rivers, deeply cut channels, huge mountain belts, wave-cut shorelines, or sand dunes on any asteroids. But the bodies do hold their own characteristics, morphologic features that show us the possible long histories of these small members of the solar system.

As is to be expected, we interpret the morphology of the planets and their satellites, and the smaller bodies, by what we experience on Earth. It is not that we are geocentric, but we have no hands-on experiences in the vastness of our solar system. True, we can view the other members of our solar system from afar, using images to understand, but our explanations are only as good as the interpretations of our own world.

With that in mind, scientists have identified many features on various asteroids, and as to be expected, most of them are the result of impacts. This information derives from Earth-based telescopic studies and especially from images taken of the two asteroids Gaspra and Ida as the Galileo spacecraft traveled to Jupiter:

- Like the planets and satellites of the solar system, craters caused by impacts seem to be the most prevalent features on the asteroids. Gaspra and Ida are pitted with large and small craters, evidence of the multitude of impacts that have occurred on the minor planets. In fact, some scientists believe that even though many impact craters pepper the surface, there are not as many as anticipated, and this may indicate a relatively young surface on both asteroids of at least one billion years old.

- Dozens of huge boulders are found on both Gaspra and Ida. For example, the image resolution for Ida revealed boulders ranging from 30 to 150 meters in diameter. In addition, regolith, a thin, soil-like covering, is associated with the Earth's Moon, but it is also apparently found on several asteroids. These features are the most surprising found on the asteroids to date: After all, the tiny bodies have little in the way of gravitational pull (Gaspra's escape velocity is only 10 meters per second, with even the gentlest impact sending the debris flying into space), and it seems illogical that the material ejected from an impact, either in the form of a boulder or regolith, would settle back on the asteroid. But apparently, the layer of pulverized rock was able to form a thin blanket on some minor planets. Boulders and regolith on the surface may not be asteroid ejecta as such, but it may have been gathered up by the asteroid as it continued in its orbit after an impact.

- Cracks, or fractures, in the surface of the Earth are usually associated with faults—long, mostly linear features representing weakened parts of the Earth's surface, split apart most often by the natural movements within the planetary crust. Cracks on the asteroids represent something much more catastrophic: Most of the splits are from the impact of large bodies, mostly other asteroids, at some time in the bodies' histories.

The terms *grooves*, *fractures*, and *cracks* have been used interchangeably when describing certain features on the asteroids. For our purposes, cracks and fractures are small, thin lines of cracks in the asteroid surface; whereas grooves are much more pronounced ridges and valleys in the asteroid's surface. The grooves found on Gaspra and Ida are muted and give both asteroids a wrinkled appearance. The cracks are evidence of the battering that the tiny worlds have endured for hundreds of millions of years. (The Martian moon Phobos has long, nearly parallel lines that run over the surface of the moon, over and up craters, apparently covering most of the surface. Some scientists believe that Phobos was once an asteroid, but, as we will see, the idea is questioned because of the complexity of dragging such a body into Martian orbit. Scientists speculate that

Phobos' grooves may mark deep internal fractures caused when the moon's largest crater, Stickney, measuring 10 kilometers in diameter, was created.) (Figure 1)

ALL SIZES AND SHAPES

When an asteroid whizzes past the Earth, one of the first questions out of most of our mouths is, how big was it? We are all fascinated by the size, realizing that asteroids come in sizes as big as a house or the size of a small state.

But why determine the size and shape of an asteroid, short of curiosity? Because size and shape are the keys to interpreting the morphological history of the small bodies. For example, almost all asteroids have apparently been shaped by powerful collisions at various times in their existence, whereas the largest ones (mainly Ceres) are more rounded, indicative of the forces that drive all large bodies to become more rounded in shape.

Even back in the middle 1800s, the shapes of the asteroids were thought to be angular. Again, from M. O. Mitchel, in 1868, from the Dudley Observatory in Albany, New York:[1]

> When carefully watched some of them exhibit rapid changes in the intensity of their light, sometimes suddenly increasing in brightness, and again as rapidly fading out. These changes have been accounted for on the supposition that these worlds are indeed angular fragments, and that, rotating on an axis, they sometimes present large reflective surfaces, and again angular points, from whence but a small amount of light reaches the earth.

Visual observation with telescopes is not the best way to measure the size or shape of an asteroid. After all, the objects are merely tiny disks when seen through a telescope. So for years, one of the common methods of determining the size and shape of an asteroid included the occultations of stars: watching how a star disappears then reappears as an asteroid comes between an observer on Earth and the star. Gathering such data from several sites around the nighttime viewing region sometimes results in a rough outline of the asteroid. This method was also used to dis-

FIGURE 1. Phobos, one of the two moons around Mars, exhibits amazing grooved features that form parallel ridges and valleys over most of the surface. The surface is further heavily cratered and covered with loose material that has "drained" into fractures forming the grooves. These linear features may be the result of the impact that formed Phobos' huge crater Stickney. This image was taken by the Viking Orbiter 1 in 1977. (Photo courtesy of Peter Thomas)

cover the rings around the planet Uranus by researchers onboard the Kuiper Airborne Observatory in 1977. As the rings came between the observers in the plane-observatory and the star, the star faded in and out; the interpretation of the alternating dimming and brightening of the star was that the planet had rings. It was confirmed several years later as the Voyager spacecraft flew by Uranus.

One of the best methods, excluding an actual visit to an asteroid, to flesh out the size, shape, and other characteristics of asteroids is ground-based radar. Such radar observations are powerful for a number of reasons, especially the high degree of control exercised by the observer on the signal transmitted to illuminate the target. In fact, the technique is so extremely precise that the radio signal's time/frequency modulation and polarization measurements can be matched to a particular scientific objective— thus more data about a particular target asteroid is received.

Radar has been a boon for asteroidal studies. Not only does it provide information on sizes, shapes, and spin of the planetoid, but it also provides such surface information as topographic relief and, occasionally, the porosity of the regolith and metal concentrations, information that would be useful if we ever wanted to visit and mine an asteroid. In addition, many near-Earth asteroids have been "imaged" using radar. Similar to radar images showing what geologic features lie deep on the ocean floor, the space radar reveals the asteroids that whip past our planet.

The major radar sites for watching asteroids are located at the National Astronomy and Ionosphere Center's Arecibo Observatory in Puerto Rico (the facility was upgraded in 1995) and the Jet Propulsion Laboratory's Goldstone Radar in California. In a typical radar experiment, the signal is transmitted to the target asteroid for a duration that nears the round-trip light travel time to the asteroid (or the amount of time for light to travel back and forth from the Earth to the asteroid). Then the echoes are received for the same duration. The returning maser-amplified signal is mixed to lower frequencies and filtered, with the resulting data as digital samples of the signal's voltage. The radar echo strength of the asteroid, which will determine how useful a radar image will be, becomes a matter of techniques and nature: It depends on the transmitted power from the original signal, the radar wavelength transmitted, the antenna gain and effective aperture, and the distance to and cross section of the asteroid.

Additional interesting findings of shape, size, features, and composition from radar have been based on asteroid echo spectra and their interpretation. For example, an echo radar image of the large object 216 Kleopatra indicates that it is two to three times longer than it is wide. Some scientists speculate that this is because the asteroid is either dumbbell shaped or a contact binary, two asteroids that are in gentle, physical contact with one another. Asteroid 1986 DA indicates an enormous radar albedo (high reflectance of the radar signal from the asteroid), meaning that the radar-bright asteroid may be very rich in metal, with hardly any regolith covering its surface.

Another good example is asteroid 4769 Castalia (1989 PB). In August 1989, astronomer Eleanor Helin discovered the 1.5-kilometer near-Earth asteroid. Two weeks later, R. Scott Hudson of Washington State University and Steven J. Ostro of the Jet Propulsion Laboratory used radar echoes to image the asteroid when it was 5.7 million kilometers from Earth. The images reflected 2.5 hours of rotation and showed a double-lobed object. Several years later, computer reconstructions of the radar images revealed two lobes, each about 0.75 kilometers across, seemingly connected with a narrow "waist"—possibly an indication that the two lobes were once separate but came together after a gentle collision.[2] (Figure 2)

Another growing field is passive microwave observation, which looks at the actual microwave emissions from the natural surface of the asteroid, a method perfected by its use on Earth satellites and aircraft. The Very Large Array (VLA) of the National Radio Observatory in New Mexico is the major observing center for microwave observations, measuring the continuum emission of the largest asteroids. So far, the results have given us a glimpse of some of the larger asteroids' surfaces. For example, microwave emission data analyzed by W. J. Webster and K. J. Johnston indicate that Ceres has a finely divided layer about 3 centimeters deep overlying a much more compact layer, with a sharp transition

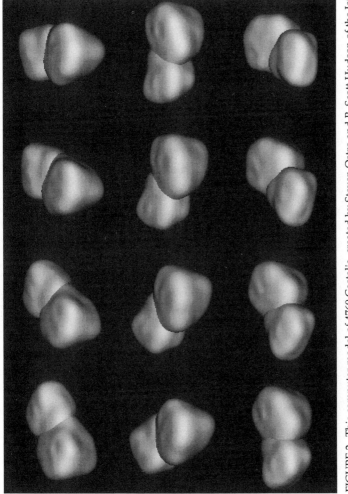

FIGURE 2. This computer model of 4769 Castalia, created by Steven Ostro and R. Scott Hudson of the Jet Propulsion Laboratory, shows 12 different views of the near-Earth asteroid. Radar data gathered at the Arecibo radar/radio telescope in Puerto Rico was used to determine the intricacies of the asteroid, gathered when the small body was 5.6 million kilometers away. It is considered the first conclusive proof that "contact binary" asteroids exist in our solar system. (Photo courtesy of Steven Ostro)

between the two. Vesta has a finely divided layer about 6 centimeters deep overlying a much more compact layer, with the surface appearing to be composed of basalt and basaltic dust. And Pallas also has a finely divided layer about 6 centimeters deep, but the subsurface is unknown.[3]

Other, more advanced techniques are used to look at asteroids: Advanced speckle interferometry is used mainly for shape and pole determinations and asteroid image reconstructions. Even radiometry, which compares observed intensities of objects at the visible and thermal-infrared wavelengths, helps to determine asteroid diameter and albedo (brightness).

Initially, we have found out more about asteroids than was ever known before. For example, we know that the largest asteroid, 1 Ceres, is about 1000 kilometers in diameter, rotates every 9 hours and 5 minutes, reflects only about 10 percent of the sunlight falling on it, and orbits the Sun every 4.61 years—meaning a Cerean "year" takes 4450 Earth-days to spin around the Sun once.

We are keeping watch on these smaller members of the solar system, learning all we can about them. Keeping an eye on the asteroids will also help us categorize the objects just in case we do increase our efforts to live in space and find a way to use the bodies to expedite our explorations (see figure 3).

SEEING DOUBLE

A common characteristic in nature is duality; features displayed by living organisms often come in pairs. There are the leaves of certain plants that grow in pairs; most animals have two eyes for depth perception; and most walking, flying, or swimming organisms have two opposed limbs in the forms of arms and legs, fins, or wings. Such doubles are often evolutionary features, a necessary symmetry to allow a creature to move, see, or function in order to survive.

Nature also apparently plays doubles with asteroids—but not for the same reasons.

The idea of binary asteroids has been around since the beginning of the century. But it was not until 1979 that anyone paid strict

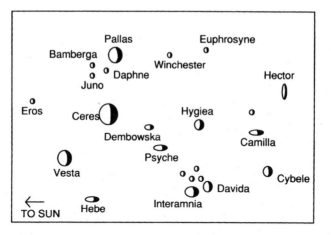

FIGURE 3. Presented here is a representation of a few larger asteroids that roam the asteroid main-belt, all in their relative position from the Sun (not to scale).

attention to this possibility. It was then that astronomer William Hartmann of the Planetary Science Institute proposed that double asteroids do exist and are probably the result of a larger asteroid's destruction. After the fragments blew clear of the catastrophic collision, they moved along nearly identical paths. As time passed, the gravity from both fragments bound them together in a feeble pairing, making the doubles, in most cases, merely siblings. Hartmann's explanation also supports the formation of contact binaries, asteroids that do not collide violently but gently nestle together to form an often dumbbell-shaped object. Experiments and theoretical models supported Hartmann's claim. And since that time, especially within the past decade with the use of radar to image the asteroids, researchers have continued to find more and more definite evidence of double asteroids.[4]

Another push to accepting binary asteroids came from asteroid occultation reports. In the late 1970s, several amateur astronomers witnessed a double dip in light during an asteroid occultation, but not everyone agreed on the results. Interestingly enough, David Dunham, who is known as the "heart and soul" of the International Occultation Timing Association, argued that amateurs had seen such doubles for years based on occultation data. There are many examples: For Texas amateur observer Paul Maley, the 1977 occultation of star Gamma Ceti A by asteroid 6 Hebe should not have disappeared for about a half second, because he was well outside the occultation track. In 1978, 532 Herculina occulted a 6th magnitude star, but several astronomers, including amateur James McMahon in California and Edward Bowell and Michael A'Hearn at Lowell Observatory in Arizona, all saw other disappearances of the star well before the main occultation. Bowell and A'Hearn reportedly remained iffy about actually perceiving a double asteroid; the problem with this particular observation has been subsequent observations of Herculina by the Hubble Space Telescope and speckle interferometry, neither of which has turned up a double.[5]

It took a close encounter with the asteroid 4769 Castalia in 1989 before double asteroids came into vogue, when Steven Ostro's radar image of the small asteroid was confirmed as a double-lobed structure, a contact binary asteroid. Then came Ostro's radar images of 4179 Toutatlis in 1992, using improved radar techniques and the new upgrades at the Arecibo radiotelescope. The asteroid had a dumbbell shape that was later interpreted as a contact binary asteroid with 1.3- and 2.7-kilometer lobes to complete the couple. Of the few imaged close encounters the Earth has had with asteroids, two were found to be potential doubles.[6] (Figure 4)

But probably the most intriguing evidence of asteroid doubles comes from the doublet craters found on the other planets and satellites of the solar system. For example, on our Moon, the craters Messier and Messier A in Mare Fecunditatis were probably formed as two asteroids struck the Moon one right after the other. Another example is the lunar crater Bessarion B in Oceanus Procellarum, a double in which the larger crater is about 12 kilometers across. The duo display a ridge (septum) of material between each crater, evidence of how the ejected material was squeezed like toothpaste from a tube at the time of their violent and almost simultaneous impacts.

Another example of doubles comes from Venus: The Ma-

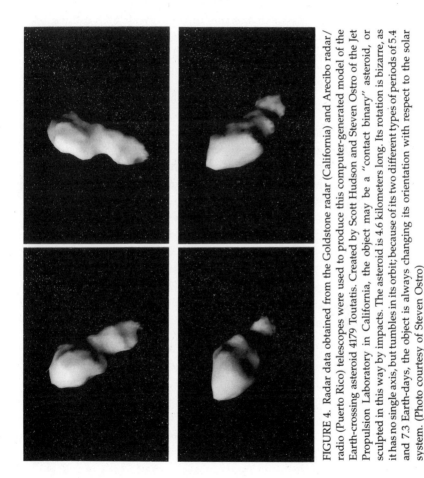

FIGURE 4. Radar data obtained from the Goldstone radar (California) and Arecibo radar/radio (Puerto Rico) telescopes were used to produce this computer-generated model of the Earth-crossing asteroid 4179 Toutatis. Created by Scott Hudson and Steven Ostro of the Jet Propulsion Laboratory in California, the object may be a "contact binary" asteroid, or sculpted in this way by impacts. The asteroid is 4.6 kilometers long. Its rotation is bizarre, as it has no single axis, but tumbles in its orbit; because of its two different types of periods of 5.4 and 7.3 Earth-days, the object is always changing its orientation with respect to the solar system. (Photo courtesy of Steven Ostro)

gellan spacecraft images revealed that about 2.5 percent of the craters on the cloud-covered planet were formed by binary impactors (there are fewer doubles there than on Earth—probably because most of the asteroid binaries disintegrate in the thick atmosphere rather than strike the surface). (Figure 5)

Even our own world has not thrown away all vestiges of double craters, and it is estimated that about 5 to 10 percent of all Earth craters are doublets. Three well-known terrestrial double craters are:

- The Ries crater in Germany, with a diameter of 24 kilometers, and its double, Steinheim, with a diameter of 3.8 kilometers, about 46 kilometers to the southwest. Both are about 15 million years old.
- The Clearwater Lakes in northern Quebec are 32 and 22 kilometers in diameter—a double crater that formed when two asteroids of about the same size smashed into the Canadian shield about 290 million years ago (see figure 6).
- The Kamensk and Gusev craters in Russia are about 25 kilometers in diameter. They are also about 65 million years old, but they are not associated with the Cretaceous–Tertiary extinctions of the same time (other impact craters in the Western Hemisphere are associated with that time of massive extinction).

By now, William Hartmann's idea of double asteroids made a great deal of sense. In fact, computer models of Hartmann's idea by such scientists as Dan Durda of the University of Arizona's Lunar and Planetary Laboratory have shown that doubles form somewhat easily. Durda's model takes a 100-kilometer parent asteroid and smashes it into thousands of pieces. He found, after the debris scatters, that there is a better than 50 percent chance that at least one of the largest fragments from the shattered asteroid will gravitationally attract a moon. Apparently, a stable orbit will result in the smaller moon orbiting the larger asteroid. An unstable orbit will result in a contact binary asteroid, both chunks of rock eventually touching as they are gravitationally attracted to each other over time.[7]

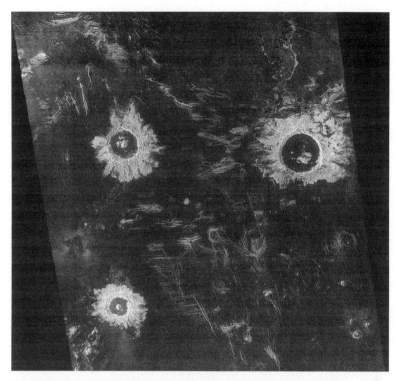

FIGURE 5. These three impact craters on Venus are in the Lavinia region of the planet. They measure from 37 to 50 kilometers in diameter. The three impact craters appear to be relatively the same age. Could they have formed when a group of asteroids or comets struck the planet almost simultaneously? Or are they only relatively close in age? (Photo courtesy of NASA)

The contemporary presence of binary asteroids may or may not be common, but another interesting note comes into the debate: Two scientists at the Lunar and Planetary Laboratory in Tucson, Arizona, H. Jay Melosh and William Bottke, have proposed that a significant fraction of the larger, mile-wide asteroids that have struck the Earth in the past were doubles. They point to other planetary bodies, especially double craters on the Moon, Mars, and Venus, as additional evidence.

Grounded in studies of dual impacts on other worlds and

FIGURE 6. The Clearwater Lakes in Quebec, Canada, represent a double strike of space bodies on the Earth. These twin impact craters formed in the Canadian shield, a place where Precambrian rocks (the oldest geologic layer of rocks on the planet) are exposed on the surface. (Space Shuttle photo courtesy of NASA)

on Earth, some scientists have tried to visualize how such an event took place on our planet. Based on a computer simulation of asteroids closely approaching the Earth, Bottke and Melosh show that the larger asteroids break into two fragments that begin to orbit each other. An additional encounter with Earth and its greater gravitational pull can cause the orbiting chunks of rock to impact the Earth.

The actual evidence from field studies of Earth impact craters shows that nearly one crater in seven that is larger than a half-kilometer across has a companion crater nearby. The breakdown of the relative ages of such impact craters show they were formed at the same time. Bottke and Melosh further contend that these impact craters could not have been formed by either the breakup of the asteroid within the planet's atmosphere or by planetary tidal forces acting on the asteroid just before striking the planet. The craters forming the doublets had to have been traveling together long before they reached the planet.[8]

Although the reason for the breakup of an asteroid near the Earth is clear—the gravitational (tidal) forces of the Earth pull the original asteroid apart—it is a puzzle why asteroids would be so fragile. One theory is that the early asteroids were much like the early planets, easily fragmented by the least gravitational or collisional provocation because they had little chance to form. Another theory states that tidal forces during the first close encounter with the Earth caused two side-by-side asteroids to separate just a little; during the next encounter with the Earth, tidal forces opened the separation of the asteroid a little more just before impact.

Such theories may also explain the doublet craters on the more barren planets and satellites of the solar system, such as the planet Mercury, or on satellites such as our Moon and many of the moons of the outer solar system. The surfaces of these bodies are essentially frozen in time, revealing the conditions that prevailed during the bombardment of the early solar system. After all, these are really "fossil" planetary surfaces, as there has never been an atmosphere to grind down and modify the surface or violent movements to churn up the crust.

But what about the more recent doublet craters? The surfaces of the Earth, Mars, Venus, or even some of the Jovian satellites show marked and geologically recent resurfacing. Yet tucked away in various parts of these planetary bodies are recent double craters from impacts that formed in pairs. Does that mean that there are more modern asteroid binaries than previously thought?

This may be true. Beside the doubles found by Ostro, more have been turning up through additional radar studies: Near-Earth asteroids 1627 Ivar and 1986 DA, the main-belt asteroid 216 Kleopatra; and Earth-approaching asteroids 433 Eros and 1620 Geographos (both show light curves that vary, which may be indicative of rotating contact binary asteroids). In addition, the small moon, Dactyl, around the main-belt asteroid Ida, was found in 1993 as the Galileo spacecraft raced past the doublet on its way to Jupiter.

As noted earlier, the number of large paired craters on Earth is about 5 to 10 percent of all impacts discovered; several scientists believe that this may represent the percentage of modern near-Earth asteroids that are doubles, too. But in reality, the actual number of near-Earth doublets is a matter of conjecture. After all, the dynamics of our planet and the 70 percent ocean cover make finding asteroid impact craters difficult, and the numbers may be much higher. For the other planets of the inner solar system, the numbers differ, too. For example, on places like Venus, the smaller asteroids, as we have noted, probably disintegrate more readily in the planet's thick atmosphere than larger asteroids, leaving fewer doubles than if there were no atmosphere.

MORE THAN A DUET

Just as nature seems to favor double features in living creatures on Earth, perhaps there is a significant number of asteroidal doubles floating around the solar system. But there are other asteroid and comet wonders that exist, including the possibility of multiple comets or asteroids, creating a chain of impacts. Could coming within the gravitational field of a planet cause the breakup or fragmentation of an asteroid or comet, creating not doubles, but multiple bodies that could eventually impact on a planetary body, like a string of pearls striking the surface?

Independently, H. Jay Melosh and Ewen A. Whitaker of the Lunar and Planetary Laboratory and Robert Wichman and Charles Wood of the University of North Dakota studied close-up photos of the lunar surface and found a crater chain candidate: a chain of impacts that stretch 47 kilometers across the Moon's crater Davy (with 23 craters) in Mare Nibium. (There is also a similar chain found near the crater Abulfeda, but it has been degraded by bombardment from other impacts over millions of years; some of Jupiter's Galilean moons also share the distinction of possessing crater chains.) According to Melosh and Whitaker, these multiple craters may have been created by a tidally disrupted "rubble pile" of asteroids. The pile was drawn into the gravitational pull of the Moon, and eventually struck en masse on the lunar surface.

I can remember years ago, when these types of multiple craters were described as originating from an asteroid or comet that came in from an oblique angle and skimmed the surface of the Moon, much like skipping a stone over water. Such impacts would produce the round craters, but nevertheless, such skipping could not really explain the neat row of holes left by the impacts. Another explanation was that they were trails of fallen ejecta from larger impacts nearby; but if crater chains formed in this manner, they would be much more common on all the planetary bodies in the solar system.

More recently, scientists believe they have discovered two such chains of craters on the Earth—although both chains still need to be confirmed. One is a line of impact craters in the African nation of Chad, formed with fragments of an asteroid or comet that slammed into the Earth about 360 million years ago. The two craters were detected around the Aorounga impact site with images taken by the space shuttle Endeavor in early 1994 (the images were from synthetic aperture radar data). The second chain of eight craters appears to stretch for about 700 kilometers from southern Illinois to Kansas.

And of course, the vindication of everyone advocating multiple craters was the recent strike of a string of objects on Jupiter. The long lineup of 22-odd fragments gave a clear indication that double, even multiple, strikes were not a figment of scientists' imaginations. This scenario was played out in full view of the

Earth: Shoemaker–Levy 9, probably a comet, broke up and eventually entered Jupiter's atmosphere in July 1994. Apparently, 2 years before, the original comet broke up on close approach to Jupiter because of the gas giant's massive tidal forces.

Still, the fragments continued in their orbit, determining their destiny. They would eventually meet again with Jupiter, but this time, they would plunge to their demise into the gas giant's atmosphere. Scientists determined the trajectories and watched. For one week, more than 22 fragments of what was Shoemaker–Levy 9 crashed into the planet's atmosphere. It was the first time humans directly witnessed such a bombardment, which provided plenty of ammunition for the researchers proposing the double and multiple asteroid theories.

The doubling or multiplying of an asteroid (or a comet, as they are known to break up into multiples, too, as we will see) is an interesting exercise in the workings of physics with the smaller bodies of the solar system. But as we explore later, such multitudes of rock pieces may pose a major problem for those of us on Earth. After all, we may find ways to destroy a comet or asteroid heading on a collision course with the Earth, but could we manage more than one at a time?

Chapter 5

COMPOSING MINOR PLANETS

What is most essential is invisible to the eye.
ST. EXUPERY
THE LITTLE PRINCE

SHOWERS FROM THE SKY

In order to understand the composition of asteroids, a planetary scientist has to be part detective. Undoubtedly, indirect clues, such as remote observations, are essential to solving the case of the asteroid's composition. A spectrographic analysis of the light reflected from the asteroid's surface, or even the changes in reflection as the asteroid spins on its axis, offer clues to its composition. But like all good detectives, the scientist must also search for direct clues in his quest for asteroidal composition. And the best such clues are meteorites—large and small, and diverse in chemical composition—that fall to the Earth.

When a meteoroid, a millimeters-sized particle or small pea-sized rock in space, enters the Earth's upper atmosphere and creates a momentary bright trail across the sky, it is called a *meteor*. Meteors were known in Aristotle's day, about 350 B.C., and their name derives from the Greek word *meteoros* or *meteora*, which roughly translated means "objects lifted high into the air." Common, and somewhat romantic, terms for meteors include *shooting star* or *falling star*, but in reality, they have nothing to do with stars.

The bright flash of a meteor is caused by its contact with millions of air molecules, heating up the meteoroid as it enters the Earth's thick atmosphere. The high temperatures cause the body to vaporize, with the resulting gases (plasma) from the meteor forming a "tail." From Earth, the meteor appears as a bright trail of light, from the incandescence of the atmosphere surrounding the object, and from the burning of the solid mass itself. Most of the meteoroids that enter the Earth's atmosphere are so small that they vaporize after a few seconds or less of intense heating, which is why most "falling stars" seem to vanish so quickly from our view.

For the average meteor to be seen from Earth, it must be between about 97 to 161 kilometers in height from the observer, as the atmosphere is too thin (and thus the air molecules farther apart) at higher altitudes to produce a drag on the meteoroid. Not every person can watch the sky constantly, plus, it is rare to glimpse a meteor during the day. Scientists have estimated that the potential number of meteors that could be seen by naked-eye observers worldwide in one day would be more than 25 million; if all those observers were to use a telescope, allowing them to detect the smaller meteoroids that hit the atmosphere and are too faint to be seen with the naked eye, the number would probably be closer to 400 billion.[1]

Not all meteors are simply minute particles committing cosmic immolation. The smallest of the meteors often end up surviving their fiery entry, as they are efficient radiators of heat (because they are small, they have a large ratio of surface area to volume; or more simply, they are not very "deep"). These tiny particles are mainly found in two places: As micrometeorites in ocean sediments, the tiny particles falling through the atmosphere, into the oceans, and finally to the floor below; and, albeit more controversial, are the wispy, colorful noctilucent clouds found between 75 to 90 kilometers in altitude—thin, wavy, low-density clouds that may form by ice coating the meteoric dust.

There are also relatively larger bodies that create even greater fireworks in the sky. These meteors are called *fireballs* and can sometimes even be heard hissing and crackling as they enter the atmosphere (the term *bolide* was once used synonymously with *fireballs*). The light from the fireballs is created much the same way as it is with smaller meteorites: As the fireball falls into the thicker, lower atmosphere, the air in front of the body compresses. The resulting gaseous envelope has a much greater diameter than the solid meteor, making the fireball appear as a teardrop with a thin tail. Many times, these bright balls of fire break into many smaller fireballs, creating an even more spectacular show.

Great quantities of meteors occasionally flash across the nighttime sky in the form of meteor showers (they can reach speeds of 10 to 40 kilometers per second). These multiple trails of light can be readily seen with the naked eye and occur with some regularity as the Earth travels through meteoroid swarms or streams. Collections of particles, called streams, are naturally released from comets and eventually extend along their orbits; when a comet eventually breaks apart (many from orbiting too close to the Sun), the material travels in a pack called a *swarm*. Because the streams and swarms are relatively in the same place in the Earth's orbit, most meteors during a shower are observed shooting out from a specific constellation, called the radiant, at a particular time of the year. Table 1 lists the annual major meteor showers.

If a meteor survives its flight through the atmosphere and strikes the surface, it is called a *meteorite*. These chunks of rock, from a meteor shower or just a random object from space, frequently fall on the Earth. But most of the meteorites are never found. The reasons are simple: Seventy percent of the Earth is covered by oceans, which swallow plenty of the meteoric evidence; many of the meteorites fall into unpopulated or underdeveloped regions, including jungles, ice sheets, and mountains; and meteorites are often misidentified as slag or other types of similar-looking terrestrial rocks.

The list of meteorite associations has grown in the past few decades as scientists discover more and more about these visitors from space. Most of the meteorites are associated with the orbits of comets, with main-belt asteroids found primarily in between the orbits of Mars and Jupiter, with Mars and the Moon, and even, some claim, associations with interstellar space.

One of the first meteorites I ever saw was in the Peabody Museum in Connecticut. It was about two feet in diameter and was a flat black color, with pits that looked as if the rock had been attacked by an ice cream scooper. In reality, most meteorites are small, measuring only a few centimeters in diameter, or about the

Table 1. Annual Major Meteor Showers

Dates	Name	Constellation (radiant)	Number per hour at maximum
Jan. 1–6	Quadrantids	Bootes	110
March 9–12	Zeta Bootids	Bootes	10
April 19–24	April Lyrids	Hercules	12
May 1–8	Eta Aquarids	Aquarius	20
(associated with Comet Halley)			
June 10–21	June Lyrids	Lyra	12
June 17–26	Ophiuchids	Ophiuchus	15
July 26–Aug. 15	Capricornids	Capricornus	6
July 15–Aug. 15	Delta Aquarids	Aquarius	35
July 15–Aug. 25	Alpha Capricornids	Capricornus	8
July 15–Aug. 25	Iota Aquarids	Aquarius	6
July 25–Aug. 18	Perseids	Cassiopeia	68
(associated with comet Swift-Tuttle)			
Aug. 19–22	Kappa Cygnids	Cygnus	4
Sept. 7–15	Beta Cassiopeids	Cassiopeia	10
Oct. 10	Draconids	Draco	variable
Oct. 16–26	Orionids	Orion	30
(associated with Comet Halley)			
Oct. 20–Nov. 30	Taurids	Taurus	12
(associated with Comet Encke)			
Nov. 7–11	Cepheids	Cassiopeia	8
Nov. 15–19	Leonids	Leo	10
(associated with Comet Tempel-I)			
Nov. 15–Dec. 6	Andromedids	Cassiopeia	variable
Dec. 7–15	Geminids	Gemini	58
(associated with Asteroid 3200)			
Dec. 17–24	Ursids	Ursa Minor	6
(associated with Comet Tuttle)			

size of the average button. The larger, albeit rarer, ones can be meters in diameter. The largest meteorites found to date are the Hoba West that landed in Grootfontein, Namibia, Africa, an approximately 60-ton iron meteorite; the 34-ton Ahnighito was found in Cape York, Greenland, and is displayed at the Hayden Planetarium in New York; and the 20-ton Bacubirito was found in Sinaloa, Mexico, and is exhibited at the School of Mines in Mexico City.

Scientists estimate that 50 tons of meteoric mass enter the

Earth's atmosphere every day. They also estimate that a fist-sized meteorite strikes the Earth's surface once every 2 hours, but almost all of them go unnoticed (most strike in the oceans, others in desolate locations). In 1988, two Texas geologists claimed that the rate at which meteorites strike the Earth was 124 times higher than the estimated rate of once a year for every million square kilometers. Since that time, estimates again moved into the lower range.[2] One study, in 1990, was the now-discontinued Meteorite Observation and Recovery Project in western Canada. A network of all-sky cameras recorded the meteor falls to Earth, concluding that about nine falls yield at least a kilogram of meteorites every year in every million square kilometers, or, again, one fall somewhere in the world every 2 hours.

Scientists report that about one-tenth of the meteoric material reaches land or oceans. If this is accurate, just imagine how much mass actually burns up in the atmosphere. In fact, if the interplanetary material entering the atmosphere every year is about 35,000 to 100,000 tons, the Earth is attracting a great deal of material. Why isn't the planet getting larger, or at least heavier? Generally, the numbers are far less than they seem, and the Earth is large enough to hold the dust and debris that space throws at us, as long as the chunks are not too large. In actuality, most of these amounts are merely estimates, as no one really knows the precise amounts of meteoric mass and/or dust that fall to the Earth.

The ability of a meteorite to survive the intense heating from their trip through the Earth's atmosphere depends on a complex array of factors. Generally, the equation includes the initial velocity and mass, the angle in which the object enters the atmosphere, and the meteor's size and shape. But certain factors seem obvious; for example, if most meteoroids entered the atmosphere at a low, shallow angle, they would have to go through more of the atmosphere, and thus, more heating and material loss. And with most meteoroids, the lower the speed upon entry of the atmosphere, the less material will be lost, as the heating does not sear away as much mass.

The number of known meteorite types has increased in the past two decades, mainly because of more meteorite discoveries and better instruments to measure the rocks' distinct attributes. It should be noted here that not all minerals found on the Earth are found in meteorites. For example, feldspar, of which plagioclase (sodium and calcium aluminum silicates) makes up almost 40 percent of the crustal rock, is found in calcium-rich achondrites only; orthoclase feldspar (potassium aluminum silicate) makes up about 12 percent of the Earth's crust, and is absent in any meteorites found to date. Pyroxenes (silicates of iron, magnesium, and calcium) is found in all stony meteorites and mesosiderites; olivine (a silicate with iron and calcium) is found in all chondrites. Conversely, alloys of iron, nickel, and cobalt are common in all meteorites, and rare in Earth rocks.

Generally, meteorites are classified according to structure and mineralogy:[3]

- *Siderites or iron meteorites,* or meteorites made of the metal iron, are thought to originate from a parent body's core (to compare, most terrestrial or Earthlike planets probably formed a core of iron during their formative stages). The largest meteorites are made of iron—as they are more resistant to mass loss as they enter the Earth's atmosphere. On Earth, there are few deposits of nonoxidized iron similar to the siderites, as the atmosphere's oxygen causes the iron deposits to weather quickly by chemical oxidation (the closest rock is called Josephinite, an iron mixed with small amounts of nickel). Currently, there are 12 groups of irons based on the chemical concentration of trace elements (called siderophiles, or metal-loving elements that readily join with metals). But most people still recognize the three basic groups based on the irons' internal crystalline structures: *hexahedrites* that contain the nickel-iron alloy kamacite; *octahedrites* that contain low-nickel kamacite and high-nickel taenite (the well-known iron meteorite acid-etched patterns, the Widmanstätten figures, are in this group); and *ataxites,* irons that do not show any structure whatsoever.

- *Siderolites or stony-iron meteorites* often survive the plunge and are thought to originate from the parent body's mantle (or layer between the core and the crust of a planetary body). As the name implies, the stony-irons are a mix of silicates and iron; only about 3 percent of all known meteorites found on Earth are of this type. The four major classifications contain nickel-iron plus the listed silicate minerals: *siderophyre* (orthopyroxene; to date, only one such meteorite exists and its distinction as a stony-iron is highly debated); *mesosiderite* (calcium pyroxene and plagioclase); *pallasite* (olivine); and *lodranite* (pyroxene and olivine).

- The most commonly found meteorites are called *stony meteorites* (also *aerolites*), composed mostly of silicates and thought to originate from the parent body's outer crust. They are the most similar to our own Earth's crustal material, but still have genuine distinct elements that mark them as extraterrestrial. The two major types of stony meteorites are:

  - *Chondrites*, which include about 80 to 85 percent of all meteorites found on the Earth. The chondrites get their name from *chondrules*, the small spherical, silicate inclusions that dot a cut-away surface of the stone and are not found in any known terrestrial rocks. The actual mode of chondrule formation is unknown. There are seven types of chondrites, each distinguished by their chondrule abundance and by their chemical type (chemical types are divided into *L-chondrites* with low iron [the chondrite has a good deal of iron, but little free iron]; *H-chondrites* with high iron; *LL-chondrites* with low iron or any other metal; and *E-chondrites*, which probably formed in an oxygen-depleted environment). One of the most interesting chondrites are the *carbonaceous chondrites*. They are thought to be the most primitive of the meteorites and contain organics; their composition most resembles that of the Sun. The most important feature of these chondrites is the presence of water-bearing minerals, such as serpentine, which forms the bulk of many of the carbonaceous chondrites. They are divided into the CM , CI, CV (also includes the CAIs, peppered with calcium-aluminum inclusions), and CO carbonaceous chondrites, which represent the distinct chemical compositions between the meteorites based mainly on the percent of water, carbon, and silicates.

- The second type of stony meteorites are the *achondrites*, which are the closest to terrestrial igneous rocks than any other type of meteorite. Once thought to be different from the chondrites by their lack of chondrules (chondrules have since been found in certain achondrites), in general they are now distinguished by their origins (chondrites formed as the chondrules gathered together to form one body; achondrites formed from the melting and recrystallization of a parent chondritic body), composition, and coarseness (achondrites are much coarser than chondrites). They are divided into two main groups, based on the presence of calcium; others divide achondrites based on more detailed compositional characteristics: aubrites, diogenites, eucrites, ureilites, and howardites (see figure 1).

- We should also add micrometeorites, the smallest of the micrometeoroids that survive the trip through the Earth's atmosphere, falling slowly, sometimes taking years, to the surface. They are thought to originate from the breakup of meteors. These somewhat suspended particles often become coated with ices in the upper atmosphere, producing what are called noctilucent clouds or salmon clouds—bluish to almost iridescent clouds at altitudes between 75 and 90 kilometers. They are the highest-known clouds in our atmosphere and are seen north of the Arctic Circle during the winter.

- Perhaps we should add interplanetary dust to this list, which is often used synonymously with the word *micrometeorite* (as you will see, there is a definite crossover in terminology). The dust does not fall to the Earth like a

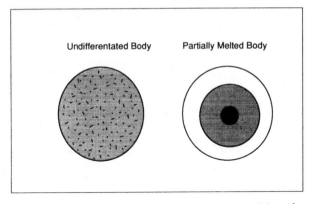

FIGURE 1. Based on their composition, meteorites appear to originate from certain layers of the parent asteroid. For example, the first parent body (left) is a chondritic mix of metal and sulfide grains, and minerals with low melting temperatures and abundant in magnesium and iron silicates, all of which appears relatively uniform throughout the body. When the body (right) becomes more differentiated from heating, it causes the metals and sulfides to form cores (the possible source of iron meteorites), and the silicates to produce low-density magmas that are vented to the surface (the possible source of calcium-rich achondrite meteorites).

meteorite but can eventually "float" to the surface through the atmosphere. It is composed of micrometeoric debris leftover from the original solar nebula, dust scattered by comets, and granular powder from asteroid and meteoroid collisions.

Comets are thought to be the culprits in producing some interplanetary dust, because much of a comet's tail is composed of dust. In the case of asteroids, the small dust particles occur when the asteroids collide, the dust and debris spraying out, similar to hitting a rock with a rock hammer. Not all the debris from the chipped rock falls into the hands of the waiting geologist—the dust usually flies from the impact point. When two asteroids collide, the effect is similar, with the small particles spreading out in all directions.

The amount of this material that falls into the Earth's atmosphere is immense. It is estimated that between

35,000 and 100,000 tons rain down on the planet every year. The dust is bound to end up somewhere below. And since the planet is more than 70 percent ocean, there is a good chance that the space-borne dust particles will end up on the ocean floors. More than a century ago, these particles were collected from the ocean, then separated out: There were those that proved to be magnetic (they contained magnetite); others were composed of tightly packed crystals of olivine and glass; still other particles were similar to carbonaceous chondrite meteorites.

The remarkable and tiniest particles (generally measuring less than 120 micrometers) are called Brownlee particles, after astronomer Donald E. Brownlee of the University of Washington, an expert on the space dust particles, and are found not only in the oceans but also in the upper atmosphere. (The smaller particles have been collected for research purposes for years; researchers request samples of certain particles by looking through the Cosmic Dust Catalogue, the scanning electron microscopic photos of each of the samples in storage. The samples are housed at the Lyndon B. Johnson Space Center in Houston, Texas, where the Moon rocks from the Apollo missions are kept.)

The interplanetary dust is best observed in the northern hemisphere in the form of zodiacal light. It is seen along the plane of the ecliptic (the path along which most of the planets travel), in the west after sundown in the early spring and in the east just before sunrise in the late fall. The faint, triangular band of light is caused by the forward-scattering of light caused when the particles come between the Earth and the Sun.

Still another type of light from the interplanetary dust is created in the eastern sky, but is much fainter than zodiacal light. The sunlight also scatters another form of brightening in the sky called Gegenschein, which resembles a faint extension of the zodiacal light along the ecliptic; it actually back-scatters the light, or reflects light from the Sun back toward observers on Earth. Comets' contributions to zodiacal light include the periodic comets, whose orbits stay close to the plane of the planets' orbits (plane of the ecliptic). Other comets with more eccentric orbits would not contribute as much dust to the zodiacal light, as their orbits take them out of the ecliptic.

One of the first methods of collecting interplanetary dust occurred in 1970, when the particles were collected by balloon. Next came the NASA high-flying U-2 aircraft in 1974, which collected the dust particles that rained down from the upper atmosphere from space. The dust and debris, caught in the Earth's gravity and slowly drifting downward in the atmosphere, collects on special collectors laced with silicon oil carried by the plane. Not that the interplanetary dust is only found around the Earth; it is also evenly distributed throughout the solar system, too. In August 1995, NASA and its German partners who ran the Galileo spacecraft on its way to Jupiter found the spacecraft plowing through the densest interplanetary dust storm ever recorded.

A HIT IN THE HEAD?

Meteorites have had their share of problems. Primarily, they usually never make landfall in one piece. But when the occasional one does slip through the system, should we be concerned?

It is probably safe to say at this point that there is an astronomical chance that someone will be struck by such a meteorite, iron or otherwise. Not that people haven't tried to explain away a few deaths in this way. One such story of a strike and kill was related in an article in the *Paducah Daily News* (Kentucky) for January 23, 1879, in which a gentleman named Leonidas Grover, who lived near Newtown, Fountain County, Indiana, was killed by a meteorite. The meteorite supposedly went through the roof, then into the chest of the unfortunate man, with the "20 pound and a few ounces" stone found at a depth of nearly 5 feet. Five feet into the ground after impacting several objects is suspicious, and like many events in older records, the entire account is suspicious.

According to Kevin K. Yau, Paul R. Weissman, and Donald K. Yeomans at the Jet Propulsion Laboratory, the Chinese have by far the most claims of "meteorite kills," citing some seven circumstances that resulted in fatalities over a period of 13 centuries. One such claim reports a rain of meteorites in 1490 over Ch'ing-yang of the Shansi Province that killed more than 10,000 people (other records indicate such a shower of meteors, but there is no mention of fatalities). Another report cites an entire family wiped out in September 1907.[4]

But there definitely have been verified near misses, some in the past and many relatively recent. Karl Jonasson, a 19th-century Swedish farmer plowing his field, narrowly escaped death when a falling meteorite landed near him; the meteorite now tops his tombstone.[5] Two young Indiana boys were standing outside on their suburban lawn at about 7 P.M. in August of 1991, when they heard a low-pitched whistle followed by a thud; there sat a warm, fist-sized chondrite meteorite in a crater about 9 centimeters wide and 4 centimeters deep. And in 1992, a woman's car trunk was hit by a 27-pound chondrite meteorite in Peekskill, New York— another near miss. She was not in the vehicle at the time.

ASTEROID FINGERPRINTS

Of the meteorites found, scientists agree that, besides those associated with comets, many originated in the asteroid belt (though recently, as we shall see, several meteorites originated from the Moon or Mars). The meteorite–asteroid connections appear to be many. Geochemists have noted that the noble gases (such as helium, neon, argon, and xenon) found in some meteorites resemble materials that would be found on the surface of an object about 3 astronomical units from the Sun, exactly in the region of the main asteroid belt. Another clue is the age of the meteorites: Most are about 4.55 billion years old. Scientists believe that the asteroid belt formed some 4.6 billion years ago along with the rest of the solar system, making the majority of meteorites chips off the old asteroids.

A major find in the meteorite–asteroid connection came in 1992, when Richard P. Binzel of the Massachusetts Institute of Technology and then-graduate student Shui Xu discovered a link between several samples of basaltic achondrites and the asteroids. They determined that the asteroid Vesta had the same spectra as the meteorites. But it did not make sense: Vesta was too far within the 3:1 resonance with Jupiter's orbit to throw the material into the inner solar system and toward the Earth. Plus, the meteorites would not be able to achieve escape velocity of this third-largest known asteroid, so it does not seem to be a source of the meteorites. But Binzel and Xu did not quit there. They identified 20 other asteroids with the same spectra, 12 which had orbits close to Vesta, and 8 which had orbits between the Jupiter resonance and Vesta. Binzel now believes that the smaller asteroids were once part of Vesta and now are repositories for past and future basaltic achondrite meteoroids that eventually make their way to Earth.

There is one puzzle: The most abundant meteorites found on Earth, about 80 to 85 percent, are the chondrites, primitive rocks that hold varying amounts of silicates and metallic iron-nickel. The majority of the chondrites exhibit extreme heating, though no melting. There are also chondrites that do not exhibit extreme heating and are probably from the asteroids, but from different localities within the asteroid belt. In particular, there are some chondrites that seem to show a marked alteration caused by water, which may mean certain asteroids once had wet surfaces. Currently, there may be no watery asteroid surfaces, because over the past billions of years, such volatiles were released. But there may be remnants—a recent look at the largest asteroid Ceres has uncovered possible water-ice on the surface.

If meteorites are extensions of the asteroids, the problem becomes, what would have caused the asteroids within the belt to heat up to such temperatures? Theories abound, including the intense T-Tauri winds from the early Sun, multiple huge impacts that produced intense heat upon impact, or the decay of radioactive isotopes, much like the way our own planet has released heat since its inception (see figure 2).

More than any other meteorite, the chondrites probably give the best evidence of what lies in the belt between the orbits of Mars and Jupiter. But for years, scientists were frustrated by the fact that the meteorites seem to have no close compositional match among any of the roughly 1000 such asteroids studied spectroscopically.

However, there is hope: In 1993, Richard Binzel and others may have found evidence of a chondrite connection. In particular, the researchers looked at smaller asteroids near 2.5 astronomical

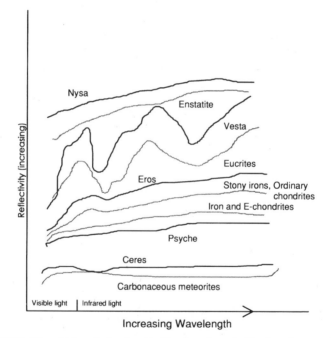

FIGURE 2. Chemical analysis of meteorites found on the Earth and spectral data from asteroids have yielded several strong connections between the two solar system members. Such information has led astronomers to conclude that the majority of meteorites are from parent asteroid bodies. The composition connection between certain meteorites and asteroids also means that we can often determine which asteroid family, or even which body, is the meteorite's parent asteroid. Meteorites are shown in gray; asteroids are shown in black.

units, which means that the objects circle the Sun in a 3:1 resonance with Jupiter's orbit (in other words, the gravitational effects of Jupiter keep these asteroids in a certain path around the Sun). The asteroids (and thus, meteoroids from asteroids) readily get thrown from this resonance lane and can end up being thrown into the inner solar system and toward the Earth. Binzel and his colleagues eventually concentrated on the asteroid 3628 Boznemcova, a mere 7 kilometers in diameter. It was apparently composed of metal-rich, calcium-poor silicates, an indication of chondrite mineralogy.

Can this one asteroid be the source of all such meteorites on the Earth? It is possible. The original asteroid could have collided with another large object, throwing chondrite material into the inner solar system, with a swarm of the asteroids coming into contact with the Earth over the past tens of millions of years.

THE NONASTEROID CONNECTION

Probably the strangest meteorites to ever plunge through the Earth's atmosphere include those that may have come from other planets and satellites of our solar system. Such finds make us realize that not all meteorites originate from asteroids or comets. Several others found on the Earth's surface may have originated on Mars, the Moon, or even interstellar space.

About 12 known meteorites are called SNCs, or Shergotty-Nakhla-Chassigny, named after their original places of discovery (more recently, many have been found in Antarctica). The meteorites are thought to have originated on Mars; the ratio between isotopes of several inert gases in the SNCs bear a striking resemblance to the ratio found by the Viking spacecraft that landed on the red planet. Scientists believe the rocks may have been ejected from the Martian surface when a large object struck. One major problem so far: No known craters on Mars—and the Viking spacecraft imaged the entire planet—are large or deep enough to support this theory.

The alleged Martian rocks also have several enigmas. One in particular is a meteorite picked up from Antarctica more than a decade ago—ALH 84001, found in the Allan Hills, supposedly from an impact that occurred on Mars about 14 to 18 million years ago. The composition of ALH 84001 is curious. It contains carbonate minerals deposited by water, with the carbonate grains banded, implying that the original rock was covered by water more than once. There are also traces of PAHs, molecules based on interconnected benzene rings, which could indicate that there was prebiotic activity on the parent planet, terrestrial contamination, or that the PAH came from the asteroid or comet that caused the impact on Mars. No one knows the real story yet.

Closer to home, scientists have determined that about 15 meteorites originated from the Moon, most with the tell-tale high ratio of iron to manganese that indicates lunar origin (two appear to be very similar, leading scientists to speculate that the rocks fell to Earth together). These lunar visitors are thought to have arrived via an impact on the Moon. There are plenty of impact craters on the lunar surface to choose from, but again, the mechanisms to send the meteoroids toward the Earth are highly debated.

When discussing Martian and lunar meteorites, scientists are confronted with the thorny problem of explaining how the rocks were blasted off the planet. We know it is possible that something this drastic seemed to have happened, since we have picked up the meteorites from the Moon and Mars. There are other clues: Near St. Gallen, Switzerland, a block of Malm limestone was discovered that apparently had been ejected from Germany's Ries impact crater (about 200 kilometers away) about 15 million years ago. Scientists know the rocks were from the impact, as the pieces exhibit shatter cones, features within the layer of rock that indicate a tremendous impact. But scientists still do not know how the rocks landed so far away, as the implied shock wave pressures were too low to propel the rock 200 kilometers.

Even the evidence of such shock wave pressures in the Moon and Mars meteorites were too low to support a blasting of rock on either planetary surface. One possible theory by H. Jay Melosh, planetary scientist at the Lunar and Planetary Lab at the University of Arizona, Tucson, asserts that it is not the pressures that are important in ejecting the debris around an impact site, but the pressure gradient.

And last but not least, there are some scientists who believe that certain meteorites may have originated from outside our solar system in interstellar space. Such meteorites are rare and are highly debated. The main problem is obvious: How would such a small chunk of rock make it through the solar system without being captured by a larger planet or pulled into orbit toward the Sun? And how, in the vastness of space, were they able to find this single speck in the galaxy called the Earth?

Chapter 6

FINDINGS FROM AFAR

The heavens themselves, the planets, and this center
Observe degree, priority, and place,
Insisture, course, proportion, season, form,
Office, and custom, in all line of order.
WILLIAM SHAKESPEARE
TROILUS AND CRESSIDA

ROCKY RELATIVES

When it comes to asteroids, holding the actual rock from the surface in your hand would be better than looking from afar. But unfortunately, there is no real way to accomplish that task at this time. And there is, without a doubt, a drawback to remotely sensed spectral data, information that we collect from Earth-based telescopes and instruments. Such data cannot accurately determine the relative abundance of the three key minerals present on asteroid surfaces, olivine, pyroxene, and nickel-iron metal. If scientists could measure the amounts of the key elements associated with such minerals (iron, silicates, and so on), then the problem of sorting through the various categories of asteroids would be easier. We may get close some day by visiting one or more via spacecraft, and then, not only would we know asteroids' true composition, but we would also solve the details concerning the connections between asteroids and meteorites.

In 1983, the Infrared Astronomical Satellite (IRAS) was launched, and observers were treated to another method, albeit space-based, that dramatically increased the ability to understand the size and shape of asteroids. Scientists knew that albedo, or the reflectance of a body, revealed a great deal about an asteroid's size and shape, but up to this time, most albedos were pure guesses. The IRAS, which conducted an all-sky survey, was able to observe the asteroids in the infrared. Although asteroids have a certain apparent magnitude from the reflection of sunlight off their surface, they also have thermal (or heat-releasing) properties. Through the thermal modeling of the various asteroids imaged, researchers developed a standardized set of albedos for asteroids and thus are now able to interpret the general composition of most asteroids (see figure 1).

Currently, asteroids are divided into numerous types, depending on their albedos (brightness), and more divisions may evolve as we discover additional asteroids. The following is a list of the most prevalent types:

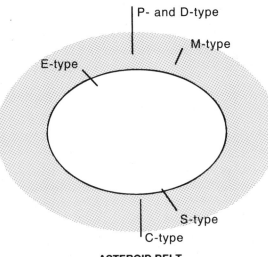

ASTEROID BELT

FIGURE 1. Certain asteroid types are thought to reside in specific orbits within the main-belt. The extent of each asteroid type is debated, although spectral data from asteroids in each range indicate that the majority of asteroids are of the types labeled in the diagram. Asteroid types are listed as letters; the lines represent the extent of each asteroid type within the belt.

- *C-types*, the majority of asteroids, are dark, carbonaceous rocks believed to be related to the carbonaceous chondrite meteorites. Their albedos range from 0.02 to 0.06, which is thought to be typical of extinct comet nuclei, and which, in turn, may comprise a large percentage of the near-Earth asteroids.

- *S-types* are grayish, stony asteroids thought to be related to the stony meteorites found on Earth. Asteroids 433 Eros [to be visited by the Near-Earth Asteroid Rendezvous (NEAR) craft toward the end of the 1990s], Gaspra, and Ida are all S-type asteroids. They are composed of iron- and magnesium-bearing silicates (pyroxene and olivine) mixed with metallic nickel and iron. There is no consensus as to whether the S-type asteroids are undifferentiated bodies related to the chondrite meteorites found on Earth or bodies similar to stony-iron meteorites that have gone through many geochemical processes of change. This also brings up a major problem in the asteroid–meteorite connection: Scientists believe that chondrite meteorites, the most common, are related to the plentiful S-type asteroids. If they are not related, where do such meteorites come from? If they are related, why do spectroscopic data from these asteroids look so different from the compositional data of chondrite meteorites?

- *M-types* of asteroids are metallic and are related to metal-rich meteorites found on Earth. Many main-belt asteroids are S- and M-type asteroids.

- *E-types* are bright asteroids made of silicate rock and chondrites.

- *R-types, also often called P- and D-types* are "red" asteroids that have reddish surfaces, probably of iron oxide; another possibility is that the red color may be evidence of certain organics. These asteroids circle outside the regular main-belt at least 5 astronomical units from the Sun (five times the average distance between the Sun and the Earth) and have albedos from 0.02 to 0.07. Many of the Trojan asteroids—those that trail or lead Jupiter in its orbit as a response to the gas giant's gravitational pull—are "red" asteroids. One red asteroid in particular may be composed of organics: 5145 Pholus, one of the most distant minor planets, that orbits the Sun in 93 years. Discovered in 1992, the asteroid was found to be redder than any known asteroid or comet. Independent teams led by Uwe Fink at the University of Arizona and Beatrice E. A. Mueller of the Kitt Peak National Observatory believe the redness is caused by a veneer of organic compounds, an idea that was proposed over a decade ago by astronomer Joseph Veverka at Cornell University. Some tests in the laboratory indicate that the spectra of organic residues, often referred to as tholins, form when mixtures of methane or other simple compounds are bombarded, in particular, by high levels of ultraviolet radiation. Thus, the residues on the asteroid may form in the same way, with the ultraviolet radiation from the Sun. The tholins' spectra in the laboratory showed steep red slopes, resembling the spectra from the surface of Pholus.[1]

- *U-types* are asteroids of unknown composition.
- Other lists include more detail: *B-type* (C-like, but brighter with a more neutral color); *F-type* (with a neutral color, flat reflectance); *G-type* (C-like, but brighter and more of an ultraviolet reflectance); *T-type* (reddish, and thought to be between a D- and S-type); and others that are S-like, but with different spectral absorptions.

FAMILY RESEMBLANCES

Held in the grasp of Jupiter and Mars, the asteroids within the main belt also exhibit a kind of continuity among themselves, largely due to the gravitational resonance of Jupiter and collisional physics of the bodies themselves. And it is probably a safe bet to say that the asteroids within these groups are somewhat similar.

This large-scale clustering of the asteroids was first noticed by Japanese astronomer Kiyotsugu Hirayama in 1918, who coined the term *families* to represent the collections of asteroids. He showed

that three groups of known asteroids have similar orbital characteristics and believed that they were all members of a single body
that experienced massive collisions in the past. In 1928 and 1933,
Hirayama carried the idea further, adding additional members to
the previous families. From his listings, other astronomers began
to group the asteroid clusters into their own classification of families, oftentimes not always stating the criteria for family member-
ship—much to the chagrin of other astronomers studying the
small bodies.

The modern view of asteroids includes families, but still with
no agreement on their number. Apparently, the families of asteroids were formed when certain asteroids collided. In particular, if
two similar-sized asteroids slammed into one another (and here
we are talking about asteroids tens of kilometers in size, possibly
over 90 kilometers in diameter) the result would be the fragmentation of the asteroids. But the pieces would not be thrown wildly
into the void of space. In many cases, the impact would be energetic enough to push the pieces into an orbit closely paralleling the
orbits of the original colliding asteroids. And to keep the asteroids
further in their distinct groupings was Jupiter, its gravitational
prodding keeping the families segregated into distinct belts.

Although the concept of families is not debated, the true
numbers of divisions are highly controversial, ranging from a
dozen to close to one hundred. The most popular families—at
least those most agreed-upon and studied—include the Themis,
Eos, Koronis, Maria, Flora, Eunomia, Nysa-Hertha, Vesta, Amala-
suntha, Mildred-Beer, Oppavia-Gefion, Dora, Adeona, and Goberta
families. (This listing was derived from a study that concentrated
on 4100 numbered asteroids; the asteroids were searched for clustering using a certain computer technique.)

Similar to a human family that appears to have similarities
among its members, there are some asteroid families that display
some semblance of continuity. For example, it is thought that
members of the Koronis family consist of a dense core with surrounding material and may be the result of a total breakup of a
parent body. Most of the Koronis family members are S-type
asteroids, undifferentiated bodies thought to be related to the

chondrite meteorites found on Earth or bodies similar to stony-iron meteorites that have gone through many geochemical processes of change. Other families are hotly debated, mainly because of our difficulty in interpreting the small bodies from such a great distance. Another problem is the great diversity within the families; for example, the Flora region is a very complex region of the asteroid belt, consisting of adjacent and overlapping families. As for the two asteroids seen close up, Gaspra is thought to originate from the 8 Flora family; Ida is thought to be associated with the Koronis family.

WHAT'S IN A NAME AND NUMBER

All the photos and searching has led to a plethora of asteroid discoveries. But with all the numbers, orbits, and future potential finds, professional and amateur astronomers alike need a way to keep up with the scorecard.

And there is a way. The 5000th asteroid was found in August 1987 by asteroid sleuth and astronomer Eleanor F. Helin of the Jet Propulsion Laboratory in California. The asteroid was named 5000 IAU, after the International Astronomical Union—the group responsible for approving names within the solar system.

Initially, an asteroid or comet is given a certain designation. For example, if the object is an asteroid, it is given the designation 1990 UL3. The "U" means it was discovered in the last half of October 1990; the "L3" means it was the 86th object found in that interval (UA1 follows UZ and I = J). The designation for comets differs: 1994e means it is the 5th comet found in 1994. A comet is also given a designation that represents the order in which it passed the Sun in a given year, for example, 1994 V; therefore, a comet [Comet Halley is a good example of numerous labels, such as 1986 III (1982I), 1910 II (1090c), etc.] that reappears can have an abundance of designations attached to it, depending on what order it spins around the Sun and the order in which it was first found on its trip toward the Sun.

As of 1995, the methodology became simpler, as astronomers

used the same system for both asteroid and comets, following the asteroidal method. Each object is now proceeded with a letter and a slash, indicative of a comet (C/), asteroid (A/), or periodic comet (P/)—a comet that has a predicted return. The objects are designated by year and a capital letter for the half of the month of discovery; a second capital letter will designate the order of discovery during that half month. For example, A/1996 EC indicates that the third object, an asteroid, was discovered in the first half of March. In addition, and similar to asteroids, known comets with established orbits (like Comet Halley) will be given numeral distinctions.

In order to control the overabundance of duplicate, inappropriate, or strange names that are entered into the naming bin, the International Astronomical Union is in charge of planetary nomenclature. Each of the planet and satellite features in our system is assigned a certain theme, usually based on the name of the planet or satellite. For example, in an obvious link of poetry and planets, Uranus' satellites were named after characters in Shakespeare's *The Tempest* and *A Midsummer Night's Dream*—Miranda, Ariel, Titania, and Oberon—and Alexander Pope's *The Rape of the Lock*—Ariel (a duplicate in Pope's and Shakespeare's works) and Umbriel. As the Voyager-2 discovered new satellites, Shakespeare was used again to fill in the new 10 satellites.

But it does not stop at just naming the satellites; there remain many features on the surface of a planetary body to be named. On Uranus' moons, most of the features imaged by the Voyager-2 were named also after Shakespeare's characters, with the exception of Umbriel and Ariel, which were given the names of evil or good spirits, respectively, from mythologies from around the world. Besides the actual name, features are also often labeled with certain Latin appendages that best describe them. For example, long, narrow, shallow depressions are called fossa (plural fossae); canyonlike chasma (chasmata); mountains called mons (montes); and sinuous valleys called vallis (valles). These labels are not always a detailed description of the features, nor do they indicate how the features formed. After all, not every mark on the outer satellites is understood, and such labels allow the feature's origins to be changed if more detailed data is gathered.

Keeping track of asteroidal findings is also connected to the Central Bureau of Astronomical Telegraphs (CBAT). Astronomer Brian Marsden is the head of the organization, which is part of the Harvard-Smithsonian Astrophysical Observatory. As the world-wide repository for the newest in asteroids, comets, or other objects that are discovered by observatories around the world, the CBAT disseminates information to other observatories to check out the validity of a claim—and with the advent of electronic mail, it often means in record time. Interested parties are then off to their telescopes to confirm or dismiss the various reported observations. Once the object is verified, it is given a distinct listing, based on the year and the time it was discovered. From there, the object is named by its founder; if the name is approved, it is cataloged into the listings for further study by other astronomers.

As for the naming of features, the only asteroids so far in contention are Gaspra and Ida. In 1995, craters on the surface of Gaspra were assigned their first names. Gaspra itself is named for the health resort in the Crimea of the same name; the craters on the surface of the asteroid were named after other such spas around the world, including Baden-Baden (Germany), Bath (United Kingdom), Carlsbad (Czech Republic), Saratoga (New York), and Spa (Belgium). The three named plains (regio) on the asteroid are named after the discoverer of Gaspra in 1916 (Neujmin Regio), the project planner for the Galileo mission (Dunne Regio), and the project scientist for the Galileo (Yeates Regio).

And if the prediction of most scientists comes true, we will be treated to the challenging enterprise of names. After all, there may be 10,000 asteroids to name from orbits closer to the Sun to those far out beyond the orbit of Pluto and Neptune. Right now, the IAU plans on naming the newly discovered trans-Neptunian objects (objects beyond the orbit of Neptune) with mythological names associated with the underworld if they are in orbital resonance with Neptune; if they are well beyond the planet, they will be given names associated with creation myths.

FIRST PHOTOS

Though radar images of the asteroids have filled in the blanks about asteroid shape and companions, we can also rely on recent visits to two main-belt asteroids. Our only close-up views of the two minor planets came from flybys of the Galileo satellite on its way to Jupiter: It showed close-ups of 951 Gaspra (1991) and 243 Ida and its moon Dactyl (1993).

The overall shape, position, and spin of Gaspra, one of the more modest asteroids at the inner edge of the main-belt, was determined long before the Galileo flew by on its way toward Jupiter. The asteroid, although small, was watched closely by Claudine Madras about a year before Galileo reached the minor planet. With the help of mentor Richard P. Binzel of the Massachusetts Institute of Technology, Madras—at the time, only 13 years old—observed the asteroid, not only from a department-store-bought Halleyscope back at her Massachusetts home, but also for two nights at the Lowell Observatory's 1.1-meter reflector. She then used photometric techniques to determine the asteroid's rotation rate, which led to an accurate 7.04 hours. To this day, Binzel claims that Madras did all the work, that he just guided her in her study of the asteroid. The work was not in vain: Madras' spin-rate research and determination of the asteroid's shape was just about picture perfect.[2]

Meeting up with the asteroid was a problem: Would the Galileo team point their cameras at the right part of the sky? Initially, the craft's cameras would be virtually shooting in the dark, hoping that along the way, between trajectory and the movement of the craft and asteroid, at least some of the computer pixels would reveal the asteroid. Astrometry measurements on the ground, along with four targeting images taken in September and October of the year, were the only way scientists were able to pinpoint the asteroid's true location, and they were within a few dozen kilometers. As it was, 150 images came from the shoot, with about three dozen taken with four-color filters, as the Galileo craft slipped past the asteroid at a distance of 1600 kilometers. Gaspra, an S-type asteroid (a mix of silicate materials and metal), measured about 20 by 12 by 11 kilometers in size, with an average radius of 6.1 kilometers. It also reflected about 20 percent of the sunlight striking it.

One of the more interesting features spotted on the asteroid Gaspra was the muted grooves. According to Joseph Veverka, a space scientist at Cornell University (the imaging team scientist for Galileo's mission at Gaspra), the grooves on the asteroid occur along linear and pitted depressions, typically 100 to 200 meters wide, 0.8m to 2.5 kilometers long, and 10 to 20 meters deep. Most seem to pack together in two major groups, one of which runs about parallel to the asteroid's long axis, and the other just about perpendicular. Scientists also speculate that the pits along the grooves are actually cracks in Gaspra's surface that once violently expelled volatile gases, such as methane—similar to "sandblows" on Earth, in which natural gases are squeezed out of the crust during an earthquake. As pointed out earlier in this book, Gaspra is not the only object in the solar system with such grooves. The Martian moon Phobos has long grooves, similar to those found on Gaspra. The connection may be no coincidence: Scientists believe that the Martian moons are both former asteroids and may have suffered the same types of catastrophic impacts from other bodies that caused the grooves.

Gaspra also has more than 600 craters that dot its surface and a very strange shape (when viewed down the long axis of the asteroid; it was once described by one scientist as being the shape of Opus the penguin's nose—a character from the cartoons *Bloom County* and *Outland*). Scientists speculate that its probable evolution included its being a huge chip off a larger asteroid. Apparently Gaspra, believed to be similar to most asteroids, was cut into its strange shape from powerful collisions at various times in its billions of years of history.

Probably the most interesting phenomena discovered from Galileo's meeting with Gaspra was an abrupt change in the direction of its interplanetary magnetic field, as detected with the craft's onboard magnetometer. A change in the orientation of the magnetic field—called field rotation—was detected three other times during Galileo's spin through the solar system, an occurrence usually associated with the solar wind. But when one of the field rotations occurred 1 minute before the flyby and 2 minutes afterward, with the redirection of the field toward Gaspra, scientists realized that they had probably detected a small magnetosphere, an oval magnetic field surrounding the asteroid. Like the bow of a ship, the magnetosphere of the small asteroid probably plows through the solar wind as the asteroid makes its way around the Sun.

The reason for the surprise is simple: Most members of the solar system, especially the smaller bodies, do not possess such magnetic fields. Jupiter has one of the largest magnetospheres known, and the Earth also has a relatively large field. But Venus and the Moon have none, and the Martian version of a magnetic field is negligible. Strangely enough, some irons, stony-irons, and ordinary chondrites do have the ability to become magnetized, which now may eventually lead to another "parent–offspring" connection between the meteorites that fall to Earth and the asteroids. In other words, if the asteroids do possess a magnetic field, the meteorites from such asteroids would exhibit indications of such a field, giving us another clue to their connection.

The actual conditions that brought about Gaspra's magnetic field, or even how small asteroids keep such a field, continues to be highly debated. Gaspra's spectra indicates that the asteroid is probably made of metal-rich silicates, or even pure metal. If so, the chances are that the object is a remnant of the core-mantle boundary of a forming planetesimal. This is the region of a planet that scientists believe creates a dynamo effect that produces a magnetic field and thus a magnetic imprint on the remnant known as Gaspra (see figure 2).

IMAGING IDA

A second image of an asteroid also was sent back by the Galileo spacecraft in 1993: Asteroid 243 Ida, another misshapen chunk of rock that seems very similar to Gaspra. For most people looking at the image, Ida would seem to be just another asteroid. But to planetary scientists, the differences between Gaspra and Ida

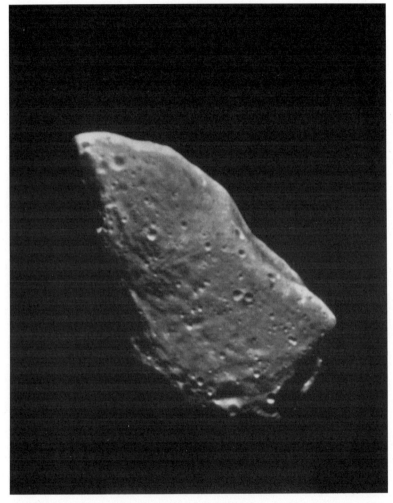

FIGURE 2. Gaspra is the first asteroid to be imaged close-up. This image was taken on October 29, 1991, by the Galileo spacecraft on its way to the planet Jupiter. The view is about 12 kilometers across; the craft was at a distance of 5400 kilometers. (Photo courtesy of Peter Thomas)

were many, especially Ida's surprise visitor that was discovered as the images were downloaded.

Similar to Gaspra, Ida is a main-belt asteroid. It is part of the Koronis family, a group of about 150 asteroids (so far) with similar orbits that may be the remains of a larger asteroid that was smashed apart during a major collision. It has a slightly reddish color, causing it to be classified as an S-type asteroid (although such asteroids are found less often at Ida's 2.86 astronomical distance from the Sun). It rotates on its axis every 4.6 hours and measures about 58 kilometers long and 23 kilometers wide.

Ida's size was quite a surprise to astronomers, who based their measurements on ground-based observations. The original estimate was a length of 30 to 35 kilometers; in reality, the cratered asteroid measured about three times the length of Gaspra. In addition, the asteroid's surface was distinctly dotted with hundreds of craters. According to astronomers such as Richard Binzel at MIT, the spin rates and shapes of the asteroids in the Koronis family indicate that the group may not be much older than one billion years (geologically, babies) and maybe even younger. The heavily cratered surface of Ida seemed to indicate a much older surface, making the asteroid's age, and the Koronis family's age, still a major puzzle.

Ida was not done. The next surprise was a distinct disturbance in the interplanetary magnetic field as the craft passed Ida at its closest spot. Analysis has shown that this asteroid may also possess a magnetic field like Gaspra, but much smaller in size. Again, the size of the object seems to indicate that a magnetic field could not be generated by the minor planet itself, but the chunk of rock carries a remnant signature of a parent body's core-mantle region (see figure 3).

Probably the biggest surprise was Ida's companion: A small moon, measuring about 1.6 by 1.2 kilometers and orbiting Ida in about one Earth-day, at a distance of about 100 kilometers. It was named Dactyl, after the entities in Greek mythology called Dactyli who lived on Mount Ida, where the infant Zeus was hidden by the nymph Ida (the Dactyli protected Zeus). The analysis of Dactyl's data indicates that although Ida and its moon are similar in color

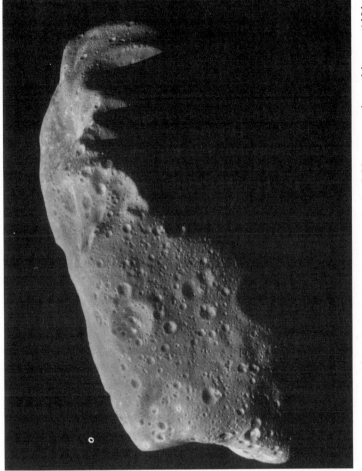

FIGURE 3. These four mosaics taken August 28, 1993, by the Galileo spacecraft are of the asteroid Ida, taken at a distance of 3000 kilometers. The asteroid is about 52 kilometers long. Overlapping craters seen in this image are 3 to 10 kilometers in diameter and create the sawtooth pattern at the top. (Photo courtesy of Peter Thomas)

and brightness, they appear to be composed of different types of material. The mineralogy of the moon (made up of almost equal amounts of olivine, orthopyroxene, and clinopyroxene) does not match its alleged parent (predominantly olivine, with a bit of orthopyroxene).

The idea of an asteroidal moon(s) has long been speculated and occasionally claimed. As with most discoveries in science, Ida and Dactyl have generated more controversy than provided answers to asteroid questions. And out of the suggestions have come several good models. One explanation of the asteroid–moon system was that Dactyl is a block of irregularly shaped rock thrown off after an impact on Ida. If that was the case, the pieces thrown out would have to keep colliding with the moon in order to shove Dactyl into asteroid orbit, a difficult hit-or-miss proposition.

Another theory (and the one more accepted) is that the two objects ended up in the same vicinity after the collision that created the Koronis family of asteroids. Apparently, two huge asteroids collided, smashing the main asteroid into pieces. For larger bodies, such impacts (and all of the planets and satellites of the solar system seem to have been affected by bombarding bodies) would not crumble the planetary body; but for a smaller asteroid, the result would be catastrophic. Over hundreds of millions of years, the majority of the smaller chunks of debris flew out of the solar system or went into independent orbits around the Sun. In this case, only Dactyl remains, gravitationally attracted and held by Ida. Of course, there is always something: The main problem is Ida's heavily cratered surface, making it appear to be much older than the alleged breakup that caused the Koronis family.

DETAILS, DETAILS

The extra bonus of finding a moon around an asteroid would lead scientists to think again about not only the possibility of double asteroids, but also how to use Dactyl to reveal more about Ida. The Galileo imaging team realized that if they could determine the size and how long it took for Dactyl to orbit Ida, an obvious application of Kepler's third law of motion would give the mass of Ida. Then dividing the mass by Ida's volume would give its overall density.

Space scientist Peter Thomas at Cornell University uses images such as those of Ida to determine the volumes of the asteroids. A body's volume is determined from stereo imaging taken as the spacecraft flies by, determining different perspectives because of its motion and the asteroid's rotation. The images are used to measure the positions of points on the asteroid surface; then a full model (latitude, longitude, and radius) is made with these points and the limbs (edges) of the asteroid. The result is a "mold" of a wire grid model that matches the available views (using a program aptly named Spud, a version of "potato" since these objects have always been called "potato-shaped"). Thomas and other members of the imaging team used views of just over a complete rotation of Ida—159 snapshots of the 58-kilometer long asteroid as it rotated for 5.7 hours (its actual rotation is 4.63 hours). Using the data to reconstruct Ida's shape, Thomas found that the resulting volume, accurate to about 10 percent, was 16,000 kilometers cubed.[3]

The orbit was more difficult to determine. The image of Dactyl is only in about 47 of the Galileo images, with the tiny moon situated about 100 kilometers from Ida and moving in the same direction as Ida's rotation. Because of this, it allowed the small moon to be easily imaged, but with the flyby path near Ida's equatorial plane, it was not easy to watch the motion of Dactyl, whose orbit is inclined by only 8.5 degrees. Taking into account the inclination and everything else, the orbit appeared to take on many shapes, and all fit the observations, ranging from 80 by 8000 kilometers to a nearly circular 82 by 95 kilometers orbit. And each orbit would mean either more or less mass for the asteroid.

The overall result was an orbit that would not allow Dactyl to fall into the asteroid any time in the near future, but it was not so outlandish an orbit as to fling Dactyl away from the lack of gravity holding it to the asteroid. Those results also showed that Ida had a density somewhere around 2.6 ± 0.5 grams per cubic centimeter. In comparison, the Sun's density is 1.4 grams per cubic centimeter and water on Earth is equal to 1 gram per cubic centimeter.

The density of an asteroid is usually considered a major link to its actual composition. But in the case of Ida, even with the density relatively pinpointed, there is still considerable debate. Is the asteroid loaded with metallic iron, as the higher density may indicate? Or is the body closer to a stony-iron or stony meteorite? Astronomer Clark R. Chapman of the Planetary Science Institute claims that Ida cannot be an overgrown stony meteorite because the density would have to be more than 5 grams per cubic centimeter. Even making Ida a traveling pile of stony meteorites would still not bring the density down to the measured level. Again, we need more data from other close asteroidal encounters to determine the actual composition of asteroids.

VESTA FROM AFAR

Recently added to the exploration of asteroids is the third-largest asteroid, Vesta. After the Hubble Space Telescope's imperfect Wide Field Planetary Camera was replaced by the space shuttle astronauts in 1993 (the initial camera was flawed during production), the planetary viewing became clearer—including our views of Vesta, the brightest asteroid.

Earth-based observers have known for years that Vesta has bright and dark markings. But the Hubble data updated the view of the asteroid. One recent interpretation of the data includes two separate sides to Vesta: One side has a spectral signature that is interpreted to be lava flows, possibly representing an ancient "crust." The other half is red, showing a signature that may represent molten (hot, liquid) rock that cooled and solidified while underground and then by some process (probably impacts) was exposed on the surface.

Whatever the morphology, scientists may have also found a link between certain meteorites found on Earth and Vesta. After all, the spectral data from Vesta reveals rocks that resemble certain achondrite meteorites found on Earth. The so-called lava flows on Vesta apparently correspond to meteorites called eucrites; the exposed former molten lava rocks resemble meteorites called diogenites. If this is true, Vesta may have been hit by large impacts in the past, throwing pieces of the asteroid fast enough to escape its gravitational pull. The chunks of Vesta may have eventually worked their way to the Earth—and may be the major source of these meteorites we find on our planet.

TRUE MOONS OR ASTEROIDS?

Perhaps more than anyone, Galileo Galilei was the catalyst for discovering moons around planets other than the Earth. In the winter of 1610, Galileo pointed one of the earliest telescopes to the nighttime sky and peered at Jupiter. There he saw four moons moving around the planet, changing their location each night. He had discovered what would later be called the Galilean satellites— Europa, Ganymede, Io, and Callisto.

The discovery of the four moons began two separate events in astronomy. First, and most obvious, the discoveries lead to the continued findings of more moons around the other planets. By 1700, five moons had been discovered around Saturn. After Sir William Hershel found the planet Uranus in 1781, he managed to find two of its moons within six years. As for the planet Neptune, which was discovered in 1851, only three weeks went by before its first moon was discovered. Second, and probably most important, the small Galilean satellites around the gas giant were strong evidence that Nicolas Copernicus and his view of the cosmos was correct: The Earth was not the center of the universe.

The spacing and number of asteroids and moons around specific planets of the solar system reveal many intriguing connections. The inner solar system has only two planets with moons, the Earth and Mars; the outer solar system has a total of probably more than 60 moons around Jupiter, Saturn, Uranus, Neptune, and the terrestrial planet Pluto. Though it does not always seem obvious, all these bodies in the solar system, from planets to satellites, were all affected by, or are directly connected to, the asteroids.

FLYING POTATOES

Imagine a moon that looks slightly like a potato with a growth problem. It is pockmarked by perhaps millions of years of impacts

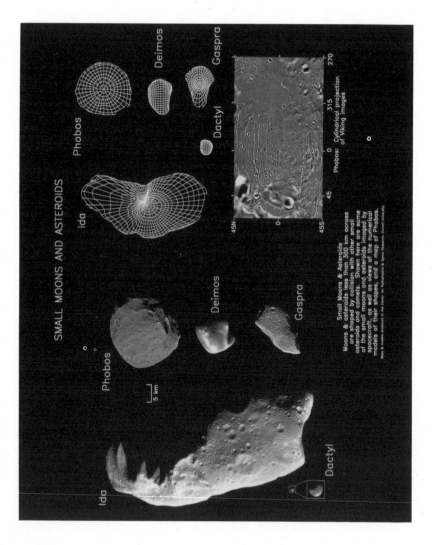

SMALL MOONS AND ASTEROIDS

Phobos

Ida

Deimos

Gaspra

Dactyl

Phobos: Cylindrical projection of Viking images

Phobos

Deimos

Gaspra

Ida

Dactyl

5 km

Small Moons & Asteroids

Moons & asteroids less than 300 km across are shaped by collision with other small asteroids and comets. Shown here are some of the small moons and asteroids imaged by spacecraft. Top row are the numerical models of their shapes, and a map of Phobos.

by debris moving through space, each projectile finding it a convenient and available target. It has bright patches that give parts of it a seemingly strange glow, probably from the pulverized rock caused by the impacts. Much of its surface resembles the lunar highlands, and there is an 11-kilometer crater on one end that would have been the death of the planetary body if it were not made of stronger stuff. Other parts of its surface are covered by a series of linear grooves believed to be the result of fractures from the tidal pull of its parent planet or even the smashing of the object that caused the huge crater on one end. It is Mars' innermost moon, Phobos (meaning "fear"), orbiting close to 5953 kilometers above the red planet. It is a moon that appears to be ready one day to crash into its gravitationally attractive master planet, but it will probably break into a ring of debris long before that occurs.

Not too far away is a much smaller, natural satellite, Deimos (meaning "terror"). This 11 by 15 kilometer, potato-shaped moon has no real distinguishing characteristics. There are no craters larger than 3 kilometers in diameter and most appear to be partially filled with regolith, essentially a thin layer of dust and dirt that covers the moon. For the visitor to Mars, the views of Phobos and Deimos moving across the sky would be much faster than that of our own Moon: Phobos orbits Mars in 7 hours and 40 minutes, rising in the west and setting in the east in 5½ hours, which is not only fast, but contrary to the usual directions of other planetary bodies. Deimos would orbit every 1¼ days and would merely be a pinpoint in the Martian nighttime sky (see figure 4).

The moons of Mars were named after the mythical sons of the Roman god of war and were discovered in 1877 by Asaph Hall,

FIGURE 4. This composite shows the relative shapes and sizes of the Martian moons Phobos (the largest moon of Mars) and Deimos; Gaspra, the first asteroid ever imaged; and Ida, with its moon Dactyl, the first asteroid moon ever imaged. Note the similarities between the moons and the asteroids—apparently old, cratered surfaces and irregular shapes—which is why many astronomers believe that the Martian moons are former asteroids. The image of Phobos was taken by the Viking Orbiter 1 spacecraft; Deimos by the Viking Orbiter 2; and Gaspra, Ida, and Dactyl by the Galileo spacecraft. (Photo composite courtesy of Peter Thomas)

when the planet was in one of its close approaches to the Earth. The importance of these two moons is twofold: First, it is illogical that these moons would have been formed at the same time as the red planet (their orbits revolve prograde to other planetary bodies and some scientists believe the moons are too small to have originated from planetary accretion around the planet when it was forming). Second, no one really knows how the moons managed to spin around Mars. Many scientists have speculated that the moons are captured asteroids from the asteroid belt; after all, Mars may have had a voracious appetite for moons and found a huge reservoir right in its backyard.

For years, most studies I encountered concerning the Martian moons stated that Phobos and Deimos *were* captured asteroids dragged from the nearby asteroid main-belt. How else could the two asteroid-looking bodies, so close to the asteroid belt, have ended up around the red planet? According to Peter Thomas, space scientist at Cornell University's Space Science Center (who has long studied the moons of Mars and, more recently, the asteroids Ida, Gaspra, and Vesta) data indicate that the moons' spectra identify them as D-type asteroids, possibly linking them to the asteroid belt. But there are problems with the theory: The D-type asteroids are found in orbits far beyond the red planet's orbit, at about five astronomical units (five times the average distance between the Sun and the Earth). Add to this the unknown composition of the D-type asteroids—a subject that is highly debated—and again, there is question of capture.

Right now, there is a standoff between "the dynamists," who are trying to determine how the asteroids could be captured, and "the spectroscopists," who explain that, based on composition, the asteroids are D-type and formed nowhere near the Martian orbit. Probably the greatest difficulty for the dynamists is explaining how the asteroids were actually captured. After all, it appears that if Mars captured an asteroid(s), it would have had to take place during the very early stages of solar system formation, when the extended protoplanet's atmosphere (or own part of the "nebula" as it was forming) could have slowed down passing objects just enough to capture them. Plus, the nebula would have had to be in a stage of collapse—if not, the nebula would have dragged the moons down to crash on the Martian surface.

OUTER MOONS

The outer solar system planets such as Jupiter, Saturn, Uranus, and Neptune are profuse with satellites. Not all are like the giant Galilean moons of Jupiter or the larger moons Titan, Saturn's largest moon, and Neptune's largest, Triton. They are smaller, misshapen chunks of rock, reminiscent of the smaller asteroids from the asteroid belt.

Were these small moons captured asteroids? Many scientists are doubtful, saying that the large and small satellites around each planet formed from the initial solar nebula. Like small solar systems of their own, the satellites coalesced from the nebula's material. As noted earlier, the whole process of planetary accretion is represented as a combination of two models—the homogeneous and heterogeneous accretion models. If the models are correct, then the smaller satellites of the outer solar system probably developed in the same place they reside today, the planets and their respective satellites forming like a miniature solar system. Most of the satellites probably lost their atmospheres as the Sun's early vigorous solar winds cleared the gases surrounding the various smaller bodies because their low gravity was unable to hold the gases.

In fact, there is other evidence for the *in situ* formation of the satellites: Most of the moons orbit their planet in the same direction as the spin of the planet (one major exception to the rule is Triton, the largest moon around Neptune, which orbits the planet in the opposite direction of the planetary spin; the satellite may have been captured as the newly formed planet's gravity pulled the moon into the Neptune system); plus, most of the moons in the outer solar system are inclined to the orbit of the planet's equator less than 1 degree. These in-line characteristics indicate that the moons have been there for quite some time—probably spinning and forming at the same time as the planet and probably not captured.

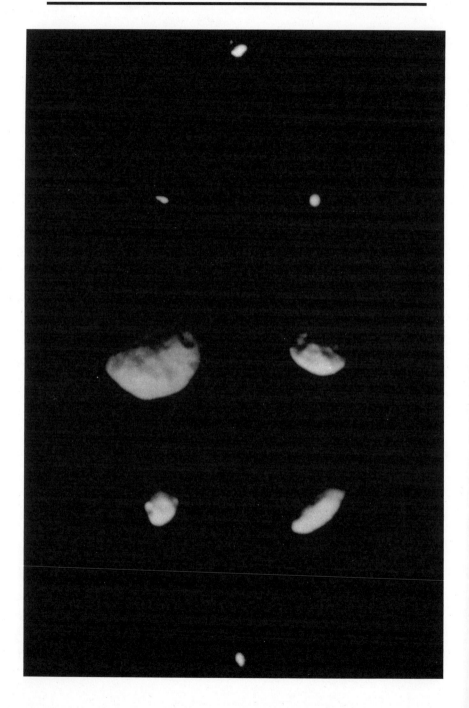

But there may be exceptions, especially among the very smallest satellites and asteroid look-alikes of the outer planets. The best evidence for believing some of the smaller satellites were captured by the respective planet is the moons' orbits and compositions. Some of the icy satellites with near-circular orbits may have been captured, as bodies (an asteroid or comet) gravitationally grabbed by a planet would tend to have such orbits due to the dynamics of capture. Composition may also be indicative of a captured moon: The rocky, carbonaceous, or icy moons may indicate either captured asteroids or the burned-out nuclei of comets.

But again, there are problems with the capture theories. Any captured asteroid is not likely to be in a very secure orbit. After all, the ability to capture implies the ability to escape. Could these moons be merely the remnants of a collision with the larger moons captured by the gravity of the planet? Or is there some mechanism that we are unaware of that controls the complex capture of the smaller bodies (see figure 5)?

Why do we care about all these asteroid details? What good is knowing about their compositions, densities, or even sizes and shapes? They are part of our solar system, and it is natural to wonder and know about our fellow inhabitants. Moreover, like some neighboring country that invades the nearby borders, one may become our enemy one day—or our friends to help us expand into space. Either way, knowing the details of as many asteroids or asteroidlike bodies as possible may help us understand how to keep a potentially destructive body away or to capture a potential resource.

FIGURE 5. Voyager imaged these smaller moons, all less than 300 kilometers, as it flew by the planet Saturn. These images revealed the moons' resemblance to asteroids: irregular shapes, battered by impacts, and very small. The smaller moons at the top "shepherding satellites." The objects in this montage are thought to have formed at the same time as the planet. Billions of years of impacts probably broke larger bodies into these fragments. Another theory is that they are captured asteroids, although they would have to have been captured relatively early after the formation of Saturn. (Photo courtesy of NASA)

Chapter 7

COMETARY COUSINS

*... the radiant star which, after traveling in its
orbit with inconceivable velocity through infinite
space, seemed suddenly—like an arrow piercing
the earth—to remain fast in one chosen spot in
the black firmament, vigorously tossing up its
tail, shining and playing with its white light and
the countless other scintillating stars.*
LEO TOLSTOY
WAR AND PEACE (ON THE GREAT COMET OF 1811)

*Lo' from the dread immensity of space,
Returning, with accelerated course,
The rushing Comet to the Sun descends;
And, as he shrinks below the shading earth
With awful train projected o're the heavens,
The guilty nations tremble.*
JAMES THOMPSON

WHY NATIONS TREMBLE

Perhaps more than any other flashy object in the sky—
even the "inconstant moon," as Shakespeare's Juliet pronounced—comets have made themselves known in the
annals of history. Cultures quivered at the ghostly specter
of the nighttime sky, the comet's plumage pointing away from the
Sun and, more often than not, coinciding with a time of great
upheaval (and, realistically, when has there not been a time of
upheaval). As we will see, for us on modern Earth, comets should
also give us pause because they have been known to come close
and even strike our planet.

Not that anyone blames the ancient nations for their collective trembling. Many of the brighter comets were (and are) awesome—fuzzy amorphous entities followed by an eerie tail, sometimes stretching across over one-quarter of the sky. It is easy to see why most people regarded the appearance of a comet as ominous. They appeared out of nowhere, and many seemed to linger in the nighttime sky for weeks. Astrologers were consulted; animals were sacrificed; and people were, to say the least, suspicious (and sometimes sacrificed, too).

No one really knew too much scientifically about these tailed wonders of the sky until astronomer Tycho Brahe realized comets were extraterrestrial objects as he tried to find the parallactic displacement of the comet of 1577 among the surrounding stars. By combining his observations with those from other observatories, Brahe determined the object was more distant than our Moon. Isaac Newton, in the next century, further demonstrated that, based on their orbital motions, comets are actually members of the solar system.

Even though more and more cometary details and orbits became predictable, the suspicions and "end-of-the-world" scenarios did not diminish. There was the comet of 1773, which was to end the world, according to Joseph-Jérome Lefrançais de Lalande. His computations of about 60 comets placed the objects close to the Earth's orbit, and although he knew a collision possibility was small, he failed to emphasize this point to a frightened public. Another comet was supposed to grace the sky, and then strike the Earth, in 1843. The Millerites believed the world would end in fire, following the locutions of William Miller, a self-proclaimed prophet whose prediction was interpreted from the Bible's eighth chapter of Daniel. (Miller was a former captain in the War of 1812; he fell from a horse-drawn wagon, hitting his head, which was suspected as contributing to his ramblings.) The Millerites predicted the end to be on April 3 of that year; when the date passed with the planet intact, the date was revised and revised again. The comet of 1843 was brilliant, but faded quickly, as did the public's interest in the Millerites and the end of the world.[1]

Amid all the shams, suspicions, and misinterpretations came

useful information, albeit indirectly: the listing of many comet passages in historical literature and documents. Armed with this information, modern astronomers have often backtracked in time and deciphered the orbital details of certain comets. Many of the determined orbital paths have led astronomers to realize that comets do have the potential to strike the Earth. Maybe the "end-of-the-world" scenarios were not so far off target after all.

COMETARY TALES

The first time I met Harvard astronomer Fred Lawrence Whipple, I was a bit embarrassed. After all, I had just handed Dr. Whipple an article I wrote for a space magazine. He was one of only 200 space scientists and astronomers chosen to be high-lighted in a special "contributors to space science" issue, and I was given a meager 200 words to write about the famous astronomer. I was extremely distressed, but such is the plight of many magazine writers as they try to whittle away enough words to explain concepts clearly to the readers, yet keep the magazine editors (and the advertisers) happy by sticking with the assigned number of words. Here was a gentleman who had a long list of accomplishments: As a graduate student at the University of California at Berkeley in 1930, he helped to calculate a rough orbit of the newly discovered planet, Pluto. He was the director at the Harvard Smithsonian Astrophysical Observatory for years, had contributed volumes to the study of comets, and so on.

Whipple is probably best known as the first person to propose an accurate description of comets. In 1951, he described a comet's nucleus as being similar to a snowball mixed with fine dust particles. His two papers detailing the orbit of Encke's Comet (with one of the shortest orbital periods known, at 3.3 years around the Sun) are still considered the most important in the history of cometary science. They show how comets, as the source of a continual supply of meteors to the solar system, were actually conglomerates of ices and meteoric particles, which aptly became known as "dirty snowballs."

The solid part of the comet, called the nucleus, is the giant snowball traveling around the Sun in a mostly elongated orbit that bears little resemblance to planetary orbits. At great distances from the Sun, the comet's nucleus, consisting of mostly water ice (although the water is not pure) and small quantities of frozen ammonia, carbon dioxide, and other exotic compounds, remains inert. It is when the body ventures toward the inner solar system that heat from the Sun begins to warm the ices and snows of the "dirty snowball," causing them to vaporize (evaporate). The released gases, along with solid particles that were imbedded in the ice, expand to form an envelope around the nucleus called the coma, or "head," of the comet. (Because of the extremely low pressures in the vacuum of space, when ice is heated, it turns directly into a gas; this bypassing of the liquid state is a process called *sublimation*.) Pushed by the solar wind away from the Sun is a distinct tail of dust and plasma (or ionized gas), streaming from the comet's head at speeds of hundreds of meters per second. The tail can be a long, thin spikelike offshoot from the comet split into several broken tails streaming out in different directions or an extremely wide, fan-shaped tail in the nighttime sky.

The only problem with Whipple's cometary definition was one of confirmation. There was (and so far, is) no way to resolve a comet's nucleus from Earth-based telescopes. As usual with most bodies in the solar system, planetary spacecraft became instrumental in our understanding of comets. In particular, in March 1986, three specially designed spacecraft flew past Comet Halley as the comet was completing another 76-year orbit around the Sun. As the comet crossed the plane of the Earth's orbit, the Soviet Vegas (two craft were sent; Vega 1 was the first to arrive), Japan's Suisei and Sakigake craft, and the European Space Agency's Giotto flew by Comet Halley, revealing the true nature of the small body, and giving us the first-ever close-up of a comet.[2]

The ability to visit Halley was due to not only having the right spacecraft technology at the right time, but also knowing the comet's orbit. Halley is a short-period comet that has been watched since around 240 B.C. (it was named after Sir Edmond Halley, the second Astronomer Royal, who first predicted the comet's return). It was also a lucky break that the comet was

Halley: Most shorter-period comets can become dull blobs of light because they produce less gas and dust from too many runs around the Sun. The many close encounters with our star creates a thick layer of dust around a short-period comet, producing a dim comet that would be difficult for a craft to image. But Halley was bright enough this time around (see figure 1).

Although the Vegas, Suisei, and Sakigake spacecraft relayed a great deal of information about the comet to a waiting audience back on Earth, some of the best and most meaningful images of the comet nucleus came from the Giotto camera: A black-and-white image of a rough, pitted body seen against the bright coma. The nucleus was elongated, measuring about 15 kilometers long by 8 kilometers wide. Parts of the nucleus had a reddish cast, probably from organics, but the heart of the comet was essentially black, with bright dust jets lit (and no doubt activated) by the heat of the Sun. Whipple's dirty snowball had been verified, although some astronomers suggest that Halley may have been more like an icy mudball. Analysis of the data from all the crafts revealed that the nucleus holds not pure water ice, but a mixture of frozen gases.

Further analysis of other comets, including those studied by chemist Clifford Matthews at the University of Chicago, reveals frozen gases of hydrogen cyanide (HCN), carbon monoxide (CO), and other ices containing carbon and sulfur.[3] If such a comet were to strike the Earth, gases from the impactor could be released into the atmosphere. But in reality, if a large comet containing these gases were to collide with the Earth, we would not be too concerned that it would poison our air. After all, the cometary gases released into the atmosphere would be relatively insignificant compared to the effects of the impact!

TRACKING POTENTIAL PROBLEMS

One of the more fascinating experiences I had in studying the stars included rubbing elbows with comet watchers. For many months during the early 1970s, at least when it was clear in the mostly cloudy northeast, a group of amateur astronomers would gather at the local observatory in upstate New York, bringing up

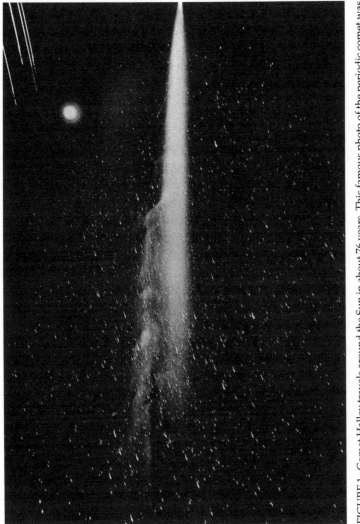

FIGURE 1. Comet Halley travels around the Sun in about 76 years. This famous photo of the periodic comet was taken in 1910. The coma and bright tail are vivid; the next visit of the comet, in 1986, was not as bright but was close enough to allow several spacecraft to visit the occasional visitor. (Photo courtesy of Lowell Observatory)

smaller scopes to spot the newest entries into the sky. The comet hunters were there nights just after the full moon, scanning the western sky (comets are found mainly when they are closest to the Sun) or they would meet at the observatory early on moonless mornings to check the eastern sky just before sunrise.

For the rest of us "noncometers," it was a pleasure watching and helping with the sky searches. It was a time when we had no charged coupled devices (CCDs; a way of electronically gathering the light from the source) to collect more light into the telescopes and no computers to watch the sky for us. We searched the sky in person, on the cold and shelterless hilltops, egging (more like begging) our motor drives to keep working. While the comet hunters worked, we would wait, using our own telescopes to single out a spot out in the sky to watch, hoping to see a galaxy we had never viewed before, to watch the nightly march of Jupiter's moons, or try to hunt down a colorful double star. And in between the sightings of those attractions came the sneak peeks at comet hunters' telescopic treats, an occasional view of a fuzzy disk hiding in a sea of stars.

I also remember it as a time when many comets were being discovered all over the world. Degree by degree, the comet watchers, professionals and backyard astronomers alike, would search the sky. For our small group of amateurs, nights were filled with striving to be the first to sight the next Comet West-type comet, or to spot the arrival of a short period comet predicted by a recent article in an astronomy magazine. And though the technology has changed, the mechanics have not: Amateurs and professionals are still finding many new comets each year. And every year, we wonder about the potential for these newly discovered comets to collide with the Earth.

No doubt, since the beginning of the solar system, icy cometary bodies have been thrown toward the planets, but humans have only been keeping track of the bodies for a few centuries. Initially, comets were classified as short-period (orbits ranging from 3.3 to 200 years) and long-period (orbits longer than 200 years). More recently, comets have been categorized into short-period, or Jupiter-family, comets (orbits less than 20 years, periodically swinging in and out of the orbital influence of the gas giant); intermediate-period comets (orbital periods between 20 and 200 years; includes Comet Halley and similar type comets); and long-period, or parabolic, comets (orbits longer than 200 years).

The major concern about these long-, intermediate-, and short-period comets is their orbits. In particular, the majority of the comets have the chance to eventually cross the Earth's orbit and perhaps encounter our world.

According to some estimates, the long-period comets that enter the inner solar system and have Earth-crossing orbits are the least likely to cause a striking problem. After all, long-period comets are on a one-way trip as they travel in their parabolic orbits. The chances of such a comet being pitched just right to hit the Earth would be astronomical. The most concern lies with the short- and intermediate-period comets. These small bodies can work their way into the inner solar system many more times than long-period comets (especially the short-period comets that are thrown around by Jupiter's gravitation), giving the objects plenty of chances to become a problem. They have the greatest potential to strike the planets and satellites of the solar system—and, of course, the Earth is included in the list of targets.

It has been estimated that only 10 percent of the known short-period comets and 50 percent of the intermediate-period comets are Earth-crossing. But these are merely guesses since the true number of comets is unknown. The numerical quandary is to be expected: Many intermediate comets have long orbital periods and have not yet returned to their closest approach to the Sun since photography was invented over 100 years ago. Plus, intermediate- and short-period comets are transient and not as easily found in the sky as more "stationary" objects, such as galaxies and clusters.

But many short-period comets are known to come close to the Earth's orbit, with the major player (again) the giant planet Jupiter. As a comet arrives in the solar system (usually by perturbations similar to those that send long-period comets from the Oort Cloud), it is often gravitationally pulled toward Jupiter. The planet kicks a comet into a path that can drastically change the small body's orbit, including into orbit close to the Earth. One example of Jupiter's power is demonstrated by Comet Lexell, a comet with a highly elliptical orbit that took it close to Jupiter. Prior to 1767, the comet's distance from the Sun was about 2.9 astronomical units. After closely approaching Jupiter in 1770, the comet was kicked into a path that took it inside the Earth orbit by 1776 and about 0.7 astronomical units from the Sun. Another close approach to Jupiter in 1779 further changed the orbit, and the comet became unobservable.

COMET RESERVOIRS

All the discoveries of comets give one pause—where do they come from? What is manufacturing these small bodies and pitching them into the inner solar system as long-, intermediate-, and short-period comets?

The best guess so far for the origin of long-period comets is the Oort Cloud, proposed in 1950 by Dutch astronomer Jan Oort. Oort suggested that a gigantic halo of comets surrounds the Sun, with a radius of perhaps 20,000 to 100,000 astronomical units. Oort's initial estimate was an average of 40,000 astronomical units from the Sun. To compare, the nearest star in the solar system, Proxima Centuri, is about 275,000 astronomical units from the Sun. Oort suggested that there must be about 100 billion comets in the cloud. More recent estimates, such as those by astronomer Paul Weissman at the Jet Propulsion Laboratory, suggest that the number of Oort Cloud members number 1 trillion comets or more, with a total mass more than 25 times that of the Earth.

The formation of the cloud of comets may have been similar to that of the planets—from the solar system's original, condensing solar nebula. Because the alleged comets formed in the outer regions of the solar system, far from the warming Sun, the cold kept the volatile gases snug within the comet. Others suggest that the Oort Cloud may not have formed at the solar system's inception, but is actually the result of perturbations from the gas giant planets. Early in the solar system's history, these planets were much gassier and larger than today, making it easy for them to push around the smaller bodies. The early comets probably formed near the outer solar system, at the edge of the condensing solar nebula. The bodies were then thrown out toward the Oort cloud region, perturbed by Uranus and Neptune; other young comets fell toward Jupiter and Saturn, and they, too, were thrown out of the solar system and into the Oort Cloud.

The main reason for Oort's proposal was to explain the existence of long-period comets. He believed, after determining the accurate orbits of numerous comets, that the cloud's innermost members were the source. These comets eventually made their way into the solar system, perturbed by passing stars or even the undulating periodic ride through the plane of the galaxy—the icy bodies losing orbital energy and falling toward the inner solar system.

Short-period comets also need a source (a long-period comet cannot become a short period comet, and vice versa), and it may be the Kuiper disk, a theoretical secondary belt of small bodies just outside the rim of the outer solar system. The idea of the Kuiper (pronounced koo-per) disk, or belt, was proposed in 1951 by Gerard Kuiper, an astronomer at the University of Chicago and University of Texas. Kuiper believed that the solar system did not stop abruptly at Pluto, but continued outward in the form of many smaller objects orbiting the Sun. As we will see, many bodies just discovered orbiting just outside the path of Neptune may be evidence of the Kuiper disk—and the origin of the short-period comets.

COMETARY CRASHES

We should face it. Comets have hit the Earth in the past and will probably do so in the future. Not all comets have such a fate, but we now have abundant evidence that some impacting comets can cause just as much of a splash on a planet as an impacting asteroid.

Perhaps the finest example of this allegory was played out by Comet Shoemaker–Levy 9, a bright string of comet fragments that struck the planet Jupiter in July 1994. Recent computer models show that Shoemaker–Levy 9 was originally a whole body just 1.5 to 2 kilometers across. Its largest fragments, which created blemishes in the gas giant's atmosphere larger than the size of the Earth, were only about 700 meters across. Its "damage" to the gas giant's atmosphere was still seen a year after the impacts, and scientists are trying to determine if the blackened areas were caused by comet debris or the dredging up of material deep in Jupiter's atmosphere. As to why such small objects could have created such major bruises in Jupiter's atmosphere, that, too, is still a matter of conjecture.

The story of the strikingly precise strike has been told many times. The initial discovery of the chain of fragments in March 1993 is credited to astronomers Eugene and Carolyn Shoemaker and David Levy. It was officially called the Period Comet Shoemaker–Levy 9 (S–L 9, or 1993e, the fifth comet discovered in 1993); the "9" means that the team had discovered 8 other periodic comets, and this was number 9 (they have also discovered more than 20 long-period comets). On March 22, the "discovery night," the team, using the Palomar (California) 18-inch Schmidt camera, took two pictures for a stereo pair more than an hour apart. The film had been somewhat exposed (someone had previously opened the box and exposed the film to light), but they decided to use the films from the center and bottom of the box with the least amount of damage. In between clouds, they took the shots, including those around the field of Jupiter.

It took until the afternoon of March 25 before the cometary fragments were found. As Carolyn Shoemaker scanned the images from the first night, she found what looked "like a squashed comet." After the initial puzzlement was over, the team reported the discovery to Brian Marsden, the director at the Center for Astronomical Telegrams (Harvard-Smithsonian Center for Astrophysics) in Massachusetts and the clearinghouse for newly discovered asteroids and comets. James V. Scotti found the comet soon afterward, confirming its presence to Marsden. The next afternoon, Marsden announced S–L 9 to the astronomical world.[4]

There was little chance that the comet would have been overlooked. Marsden explained that the broken comet would eventually have been found by others, albeit days later—even if the Shoemaker–Levy team had not chosen to use up the damaged photographic film. Five days before they took their films, astronomer Eleanor Helin at Palomar Observatory and Mats Lindgren at the European Southern Observatory independently took images of the comet in their routine sky patrol. One of them would have reported the sightings within a couple of weeks of the comet's detection. In addition, the comet was also imaged several days later by astronomer Orlando Naranjo with the Schmidt telescope at Merida, Venezuela; he reported his findings, as part of a joint collaboration with Lindgren, by March 30—five days after the Shoemaker–Levy team.

Soon, the calculations of the fragments' orbits were made forward and backward in time. Apparently, the whole body had been grasped by Jupiter's gravitational influence in 1992, then ripped apart from the tidal stresses; in the future, the fragments would collide with the planet in July 1994. And true to the predictions, on July 16, 1994, and for 6 days afterward, the fragments struck the gas giant planet.

Such strikes are apparently nothing new to the Jupiter system. Eugene Shoemaker, the codiscoverer of S-L 9, believes, based on the comet statistics on Jupiter's moons Ganymede and Callisto, that a 2-kilometer-wide comet strikes Jupiter every 2000 years; a 1.5-kilometer-wide comet strikes Jupiter about once every century; smaller comets, say, ones about 500 meters in diameter, strike about once a decade. Odds run in the favor of strikes mainly because of the huge gravitational field of the giant planet, making it a perfect target for smaller objects that get pulled into Jupiter's gravitational well.

IDENTITY CRISIS

Although most scientists believe the objects that struck Jupiter were comets, there is still a debate that occasionally rages: Were the pieces a bevy of broken comets or asteroids? The objects were first labeled a comet because the fragments all appeared to be surrounded by comas, similar to most well-behaved comets. And because it was so easily torn apart by Jupiter's strong gravitational field as it just missed the giant planet in 1992, some scientists claim this points to a loosely compacted, dirty iceball.

To an astronomer, the two bodies are distinguished by their appearance in the telescope: Asteroids are starlike in appearance, thus the name asteroid, or "starlike." If the object has a visible, gaseous atmosphere surrounding it, or has a distinct tail, often seen as a fuzzy blob around the object, it is a comet. As is to be expected, the reasons are simple: An asteroid does not have an "atmosphere," so it displays a steady light as seen through the telescope. A comet usually contains volatiles such as water-ice or gases that evaporate (especially as it approaches the warming Sun), giving the comet its soft glow from a transient atmosphere.

And indeed, one of the discoverers of the comet, David H. Levy, believes that the objects that struck the giant planet were from a fragmented comet. After all, through telescopes, each fragment appeared to be surrounded by its own spherical coma. Plus, the orbital history of the objects suggest a cometary origin. Levy also notes in his book *Impact Jupiter* that Brian Marsden, the director of the Center for Astronomical Telegrams, points out that before the gas giant Jupiter grabbed the object earlier in the century, Shoemaker–Levy 9 probably had an orbital period of about 8 years and was 4 astronomical units from the Sun (four times the distance from the Sun to the Earth).[5] Overall, most scientists agree that the objects were comets, with the original object probably beginning its life as a Jupiter-family comet.

Still, we are too far removed from the region of Jupiter to make an actual, point-blank identification. And it is true that scientists have hardly pinned down the true definitions of comets and asteroids to everyone's taste. Unfortunately, too often the determination is based on circumstantial evidence and semantics.

So what questions remain? Two observations after the impacts created doubt about the objects' identities: First, no water vapor was immediately detected (water vapor was found in the data afterward) in the plumes thrown up by the impacts, which is found in the majority of comets. Second, Fragment M did not vaporize in 1993 as predicted, meaning it did not have a coma of ice and dust from the vaporization, and thus could have been an asteroid.[5] In addition, even though the orbital history appears to be set, no one can really guarantee where Shoemaker–Levy came from—the long history of the object's orbit is too complex to trace back through time.[6]

My own problem with the fragmented group is one of size. One question that comes up over and over is why no huge dark spots have been seen on Jupiter in the past 200 years or so of observations with telescopes. Have only small chunks of comets (or asteroids) struck the planet, not even making a dent in Jupiter's atmospheric blanket? Did we witness a large, albeit crumbly, asteroid coming out of hiding from the deeper recesses of the outer solar system, only to be dragged into the gravitational well of Jupiter? Or was this a rare comet, as Eugene Shoemaker pointed out at an International Astronomical Union meeting, and, indeed, have we "been privileged to witness a bloody miracle"? (Figure 2)

Whether it was a fragmented comet or asteroid does not hide the fact that the event confirms a major theory—in this case, that a fragmented body can strike a planet and produce multiple impressions from those impacts. The black blobs on Jupiter are evidence, and as we have seen, crater chains on the Moon and the planet Mercury also confirm such multiple impacts have happened in the past.

The Jupiter impact has also increased the number of fragmented comet discoveries. In 1995, short-period Comet Machholz-2, discovered in August, 1994 (1994o), by California amateur astronomer Donald Machholz (the most prolific living comet discoverer in the western hemisphere), was five comet fragments orbiting together in a string about 1 kilometer long. Scientists believe the Earth-crossing comet, which will come within 19 million kilometers of the Earth in 2036, will eventually break into more fragments due to the effects of tidal forces.[7] And still another periodic comet was reported to have shattered, again reported in 1994. According to James V. Scotti of the Lunar and Planetary Laboratory, Comet Harrington (1994g) had split into two faint

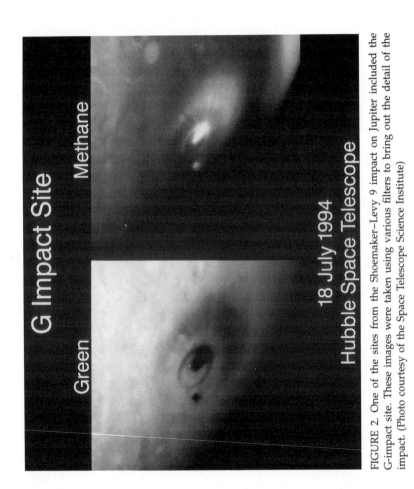

FIGURE 2. One of the sites from the Shoemaker–Levy 9 impact on Jupiter included the G-impact site. These images were taken using various filters to bring out the detail of the impact. (Photo courtesy of the Space Telescope Science Institute)

pieces (both are about 20th magnitude), a separation that may have occurred in 1987.

We have only scratched the surface of understanding the number and orbital interpretations of shattered comets, and until we gather more data, it is difficult to say that such comets could strike the Earth. But since so many fragmented short-period comets have been found and the comets in this group often become Earth-crossing comets, there is a chance that our planet has (or will eventually) experienced a cometary "shower" of fragments.

COMETS IN DISGUISE?

Interestingly enough, some exhausted comets, or those that eventually lose their volatile materials as they swing so many times around the Sun, probably appear as asteroids. The chances are high that some asteroids, especially those that swing by the Earth as near-Earth asteroids, are burned-out comets, with some astronomers estimating the number to be close to 50 percent. Several of the nearby near-Earth asteroid surfaces are similar to the dark, reddish nucleus of Comet Halley. And some asteroid orbits seem to follow eccentric paths that resemble cometary orbits, especially short-period comets.

For example, Comet Encke, a depleted short-period comet, follows an asteroidlike orbit, and in thousands of years will be indistinguishable from a near-Earth asteroid called an Apollo asteroid, complete with its volatiles depleted. Another example came in 1979, when astronomer Eleanor Helin of the Jet Propulsion Laboratory discovered asteroid 4015 (1979 VA), a typical minor planet. But when Edward Bowell of the Lowell Observatory looked through prediscovery photographic plates made from the first Palomar Sky Survey, he found a surprise: A plate exposed in November 1949 showed not an asteroid, but an object with a tail that looked like a comet. Further investigation by Brian Marsden of the Harvard Smithsonian Astrophysical Observatory revealed that asteroid 4015 was actually once identified as comet Wilson–Harrington. And in the most recent analysis, some scientists believe the object had been a comet and is now an asteroid, with the 1949 plate capturing the last gasp of the exhausted comet.

How concerned should we be about the comets' potential, either as comets or as burned-out comets, to impact on the Earth? Overall, scientists have estimated the percentage of impacts by asteroids and comets based on orbits and populations of the object, and believe that about 10 to 15 percent of all strikes on the Earth are by comets.

Such a statistic does give one pause and makes comets respectable adversaries. But with the majority of strikes caused by asteroids, most of us would rather keep a keener eye on the minor planets.

Chapter 8

WHAT ELSE IS OUT THERE?

*It is known to me that at least two American
astronomers, armed with powerful telescopes,
have been searching quite recently for a trans-
Neptunian planet. These searches have been
caused by the fact that Professor Newcomb's
tables of Uranus and Neptune already begin to
differ from observation. But are we to infer from
these errors of the planetary tables the existence
of a trans-Neptunian planet? It is possible that
such a planet may exist, but the probability is, I
think, that the differences are caused by errors in
the theories of these planets ...*

ASAPH HALL (1829–1907), PROFESSOR AND AMERICAN
ASTRONOMER, DISCOVERER OF
MARTIAN MOONS PHOBOS AND DEIMOS IN 1877.

PLANETS, BUT NOT EVERYWHERE

Contrary to popular belief, the nine planets and sundry
satellites are not the only lurkers in the region we call the
solar system. As we have discovered in previous chap-
ters, asteroids, comets, meteors, gases, and dust play
important roles in the inner and outer system. (If we wanted to be
more thorough, we could also add all the debris sent into orbit
around the Earth from unmanned and manned space missions.)
And there are more objects out beyond the farthest planet, some of
them directly tied to the asteroids.

Searching the sky far beyond our own planet has, historically,
been a matter of luck or diligent searches. The bright planets,
Mercury, Venus, Mars, Jupiter, and Saturn, were found by astute

observers of the nighttime sky from Earth. But the other planets took more time.

Uranus, the seventh planet, was discovered serendipitously by German–English astronomer William Herschel in 1781 during a sky survey of Gemini. Believing he had found a comet, Herschel followed the object for weeks before announcing the planet. Several months after its discovery, the planet's orbit was calculated: The nearly circular orbit resided beyond the orbit of Saturn. Herschel proposed the name Georgium Sidus for the planet, in honor of George III, the reigning king of England. Others wanted Herschel. But luckily for consistency's sake, the planet was named simply Uranus, father of the Titans and grandfather of Jupiter, in keeping with naming planets for gods of Greek mythology. (Actually, the planet Uranus had been "found" many times before, as had the other outer, then unknown, planets. Star charts drawn up from the 1690s to the time of Herschel's discovery had recorded the planet as a star at least 20 times.)

The eighth planet discovered was Neptune, and unlike Uranus, it was found by mathematical calculation. The first attempt at determining Uranus' orbit was accomplished by Jean-Baptiste Delambre in 1790, based on observations after the planet's discovery. But the orbit was not quite right, even after bringing in the gravitational effects of the giant planets Jupiter and Saturn. Astronomers even extrapolated back in time to the 1690s, using star charts that had labeled the then unknown planet Uranus a star. By 1840, the differences described in Uranus' calculated orbit were enough to go against Newtonian physics, amounting to differences of about 2 minutes (part of a degree in the sky)—a substantial difference when determining a planet's orbit.

The perturbations in the orbit of Uranus were further studied independently by English mathematician John Couch Adams and French mathematician Urbain Leverrier (or Le Verrier). Adams, suspecting that something was pulling Uranus out of its orbit, had sent his calculations to Sir George Airy, the Astronomer Royal of England, in October 1845, describing the location of a possible new planet. Airy did not know the young mathematician and in accordance with what seemed to be custom at that time, Airy gave Adams a simple problem to solve in order to test his mathematical prowess, but the unanswered query caused Airy to drop the matter. It was unfortunate for both Adams and Airy—Adams' calculations of the planet's location were only off by 2 degrees.

At the same time, Leverrier, unaware of Adams' work, was working on the same problem, but in Leverrier's case, the calculation came much closer, off by only one degree. Leverrier was also better than Adams at publicizing his ideas; he published his work in June 1846. Airy had Adams' calculations already in hand and wrote to Leverrier, sending the usual query—a simple mathematical problem to Leverrier to test his mathematical abilities. When Leverrier responded with the correct answers, Airy believed he had a winner, and the true calculation of a possible planet.

Airy then took Leverrier's information to Challis, director of the Cambridge Observatory, to find the alleged planet. But good star charts were hard, if not impossible, to find at this time, and the inaccuracies hindered the discovery of a planet in the predicted location (in the constellation Aquarius). And though Challis decided to painstakingly draw the sector over and over for several days, hoping to see some change in the stellar pattern that would indicate a moving object, he did not succeed. Actually, he did plot the new planet on his chart, but his methodology was not perfect, and he did not recognize the dot as a planet.

It took two other astronomers to visually discover the planet: Johannes Galle and Heinrich Ludwig D'Arrest at the Berlin Observatory. One month after Challis worked toward discovering a planet, Leverrier turned to Galle, and on September 23, 1846, requested that Galle check out the spot in Aquarius. That very night, Galle and D'Arrest found a planetary disk. It was Neptune, named for the god of the sea.

Finding additional planets beyond the orbit of Saturn gave astronomers the impetus to start looking for more. But it took another century before the last planet we have on our current list, Pluto, was found.

THE LAST PLANET

Pluto's discovery, like Neptune's, was a major endeavor, made after years of calculations and speculations. After the orbits of Uranus and Neptune were put to rest, several astronomers still believed that Neptune in particular was being perturbed just a little too much. Many astronomers, including William H. Pickering and Percival Lowell, jumped on the astronomical bandwagon, pointing to the gravitational influence by a ninth planet beyond the orbit of Neptune. (Today we realize that many of the alleged perturbations were based on errors in the original observations; in addition, as our more recent measurements of Pluto's mass continue to show it to be smaller than originally thought, astronomers realize that Pluto cannot be responsible for irregularities in Neptune's orbit.)

In fact, it was Lowell's solution to the problem of planetary perturbations that led to the actual discovery of Pluto. Lowell had predicted a planet in the constellation of Gemini, with a planet mass about 6.6 Earth masses—"Planet X," a planet beyond the orbit of Neptune. For 10 years from his Arizona Observatory (now the Lowell Observatory in Flagstaff, Arizona), Lowell checked for the unknown body until his death in 1916, but found no planet.

Lowell's dream did not die with the astronomer. He left behind three devoted assistants who would carry on his work: V. M. Slipher (director and expert in spectrographic studies), C. O. Lampland (assistant director, who supervised the Planet X searches), and E. C. Slipher (world-famous authority on Mars, who conducted visual and photographic observations of the planets, and who also dipped into politics, serving at one time as the mayor of Flagstaff and as a state senator in the Arizona legislature).

But the ability to continue the search became bogged down in estate litigation and lack of funding. Finally, with the backing of Roger Lowell Putnum, Lowell's nephew (who was eager to have the search resumed in hopes of finding "Uncle Percy's" Planet X), a 33-millimeter photographic telescope (the Lawrence Lowell Telescope, now called the Pluto Telescope) was in place at the observatory in 1929, which allowed a 12 by 14 degree section of sky to be taken in a single photograph. Again, the search began in the constellation of Gemini, but to wade through the mass of stars in the constellation (it lies near the Milky Way's spray of stars) was a big problem. But there was hope: the blink microscope similar to the one used by Max Wolf to find asteroids in the late 1800s at the advent of astrophotography.

It was astronomer Clyde Tombaugh who searched for a planet beyond Neptune in 1929 at the Lowell Observatory, under V. M. Slipher's direction. Pairs of photographic plates taken with the Lowell Telescope were then examined with the blink microscope comparator: The blink microscope rapidly alternates the views of two photographic plates taken at different times of the same star field; the moving space body would appear to "blink," or jump back and forth, against the background of fixed stars.

Pluto is the only planet to be discovered in this century, making the story of the discovery that much more exciting. Images taken with the Lowell Telescope on the nights of January 23 and 29, 1930, would be the most important. According to Tombaugh, in his book (with Patrick Moore) *Out of the Darkness*:

> On the morning of 18 February [1930], I placed the 23 January and 29 January Gem plates on the Blink-Comparator, starting on the eastern half. This was a most fortunate decision. Had it been otherwise, Pluto might not have been discovered in 1930.
>
> By four o'clock that afternoon, mountain standard time, I had covered one-fourth of the pair. After completing a horizontal strip on the left half, I rolled the horizontal carriage back to the center north-south line (which I always drew as a thin ink line on the back of the later plate of the pair). I had established this habit of progressing to the left so that I would not forget which way I was going in case of interruptions.... I raised the eyepiece assembly to the next horizontal strip. At the center line, I had the guide star Delta Gem in the small rectangular field of the eyepiece. After scanning a few fields to the left, I turned the next field into view. Suddenly I spied a fifteenth magnitude image popping out and disappearing in the rapidly alternating views. Then I spied another image doing the same thing about 3 millimeters (or .125 inches) to the left. "That's it," I exclaimed to myself. Now which image belonged to which plate? I turned off the automatic blinker and turned the shutter back and forth by a small finger lever. The right-hand image was on the earlier plate (23 January). West was to the left in the field. Then I turned the shutter to view the 29 January plate. This image was to the left of the other. Retrograde motion alright [sic]! If the direction of shift had been to the east, then the images would have been either

spurious, or were those of two independent eclipsing variable stars which happened to be caught in alternate phases of variation. Considering the interval between the plates, the parallactic shift indicated that the object was far beyond the orbit of Neptune, perhaps a thousand million miles beyond.[1]

The object was about the right distance for a planet far beyond the orbit of Neptune. The discovery was announced on March 13, 1930. The new approximately 2280-kilometer planet (the most recent estimates) was named Pluto, for the god of the underworld. And as Tombaugh explains, astronomers all over the world began to dig back into their old plate files for images that might reveal the tiny planet. The images from other observatories led to a more reliable orbit. In fact, the orbits were "in remarkable agreement with the ones predicted by [Percival] Lowell and one by [William H.] Pickering before he revised it to a poorer one."[2]

PLUTO: THE ULTIMATE ASTEROID?

Since its discovery in 1930, Pluto has always been the odd one in the solar system family. Its orbital inclination is a steep 17 degrees, and its eccentricity of 0.25 means that at its perihelion (or the closest point in its orbit to the Sun), Pluto is closer to the Sun than Neptune (it has actually been skirting inside of Neptune's orbit from 1976 and will return to beyond Neptune's orbit in 1999). In addition, Pluto has a moon, Charon, which seems to be somewhat too large to be around such a small planet.

Astronomers have speculated that Pluto actually started out in a reasonable, low inclination and circular orbit that eventually evolved into its present state due to the influence of the giant planets, especially Neptune. There are many theories, including one by Renu Malhotra of the Lunar and Planetary Institute, in which Neptune's encounters with early planetesimals gave the giant planet a great deal of angular momentum, thus sending its orbit outward by some 5 astronomical units from the Sun. Pluto was also forming at the same time as Neptune, eventually traveling into Neptune's resonance zone and, like Jupiter splitting open gaps in the asteroid belt, Neptune perturbed Pluto into an eccentric orbit. Another theory by Harold F. Levison and Alan Stern of the Southwest Research Institute states that a huge object (no doubt an early planetesimal left over from the formation of the solar system) collided with Pluto as it was in its original, more "normal" orbit. The crashing blow knocked the planet just right, sending it into an eccentric orbit, and into orbital resonance with Neptune.

But more than any other puzzle that Pluto poses, the major one is Charon, Pluto's satellite. The small moon was discovered in 1978 by astronomer James Christy and measures an estimated 1270 kilometers in diameter, compared to Pluto's 2280-kilometer diameter (with an uncertainty of 1 percent). What makes the moon so remarkable is not only its large size when compared to its parent planet, but also its comparative mass and density. As to Charon's mass, some estimates range as high as 16 percent of Pluto's mass down to about 8 percent; Charon's density is also debated, but most argue that the density ranges from 1.3 to 2.0 (Pluto's density is 2.0) (see figure 1).

The origin of the small moon is also highly debated. Many scientists lean in the direction of a captured minor planet, Pluto catching the unsuspecting object during its formation early in its history. Another theory states that Charon is a recently (geologically speaking) captured object, perhaps from the Kuiper disk or a stray asteroid, whose orbit has yet to become circular around Pluto. Another idea is that the moon is a remnant of a collision that occurred long ago, just after the formation of the planet, similar to theories of our Moon resulting from a planetestimal collision. (Interestingly enough, recent Hubble Space Telescope data indicates that Charon moves in an elliptical orbit around Pluto, an orbit thought to have been caused by an object slamming into Pluto or Charon within the past 10 million years.)

But if you like to see fur and feathers fly, mention the idea that Pluto may not really be a planet at all. In reality, the smallest planet in our solar system is teetering on the brink of planetary status. First, Pluto is too small compared to the other planets in the solar system; but it is still too large to be classified as an asteroid (or comet). Second, one of the main reasons why Pluto was called a

FIGURE 1. Pluto, the smallest planet yet discovered in the solar system, has a small moon called Charon. This Hubble Space Telescope image reveals little about the planet and its small moon. Further investigation of the data has revealed a bright spot on the planet that is thought to be a reflection off a smooth surface. Charon is somewhat bluer than Pluto, suggesting that their surfaces are of different compositions. (Photo courtesy of the Space Telescope Science Institute)

planet was because of the influence of Percival Lowell and his search for Planet X. Even though the planet Lowell sought should have been much more massive, Pluto worked its way into the literature as our ninth planet before astronomers could propose that the object be called a minor planet. Third, the motions of the planet are more similar to that of an asteroid or comet, with an eccentricity and high orbital inclination. And perhaps the most interesting idea is that the Pluto–Charon duet may be more similar to binary asteroids, objects that astronomers have discovered are much more prevalent than previously thought.

On the side of keeping Pluto as a planet, scientists point to the major attributes of all the planets. First, Pluto does orbit a major star. Second, Pluto is apparently round, not always a common feature among the asteroids (in fact, Ceres, the largest asteroid, may be the only real round asteroid because of its immense size). Third, Pluto does have a discernible atmosphere, something no other known asteroid is known to possess. Fourth, and probably most important, just think of all the textbooks and years of training we would have to change!

The only way to settle the debate may well be to visit the last planet or to continue to study the latest trans-Neptunian objects that orbit beyond the orbit of Neptune. Whatever the way, Pluto may turn out to be one of the fine lines between a comet and an asteroid, one of the largest ever to be found—or it will continue to remain our ninth planet.[3]

IN SEARCH OF PLANET X

Clyde Tombaugh's discovery of Pluto left the astronomical world hopeful for another planet beyond the small globe. Thus began the search for a tenth planet, also labeled "Planet X," a body that astronomers hoped would clarify the discrepancies in certain orbital calculations of the outer planets.

Tombaugh did not stop at Pluto. Two years after discovering the small planet, Tombaugh, instructed by the observatory director V. M. Slipher, continued the search for other objects beyond Neptune. According to Tombaugh, he was extremely thorough in the search for the possible planet:

> I spent some 7,000 hours blinking plates over a period of 14 years, and I was conscious of seeing every one of the 90 million star images! Now, if anyone thinks I might have missed seeing a planet, he or she is welcome to reblink my plates. All 362 of them! I covered two thirds of the entire sky, mostly from declination 60 degrees north to 50 degrees south. I even scanned over Canopus and the globular cluster Omega Centauri....[4]

Tombaugh, now Emeritus Professor of Astronomy at New Mexico State University, says the search for Planet X beyond Pluto is over. His exhaustive survey showed nothing, not even an in-kling of another larger body in the stellar images. Many other astronomers besides Tombaugh searched for Planet X beyond Neptune, but no one has ever found any evidence. In addition, in 1993, E. Myles Standish of the Jet Propulsion Laboratory, deter-mined that although Uranus continues to be pulled off its pre-dicted position by close to a half arc second, it is not because of a large Planet X on the outskirts of the solar system. Standish deter-mined that the actual anomaly disappears when the more accurate masses for Jupiter, Saturn, Uranus, and Neptune derived from the Voyager flybys are inserted into the equation. Another blow for Planet X.

But there is something out there. The next worlds we recently discovered on the edges of the solar system may not be so large as the alleged Planet X, but are probably closer to asteroids and comets than we know. And they are just beyond the rim of the solar system.

BEYOND THE RIM

Writers of the science fiction genre enjoy creating stories that deal with "outer space" beyond our solar system. We are strangely intrigued by the thought that there is something out there that we cannot see from Earth, like the shark lurking in the murky ocean waters just beyond our view. We grasp at things that we cannot understand and often make up grand stories about what waits for us in space if we could just get our engines to accelerate close to light speeds and beyond the confines of the solar system.

And sometimes, too, astronomers come up with what is called a theory, which to many of us sounds like a bizarre story. Take, for example, a huge halo of comets (and perhaps asteroids, or cross-breeds of comets and asteroids) in a massive ring around the solar system beyond the orbit of Neptune: This is what astron-omers call the Kuiper disk or belt.

This is not the famous Oort cloud of comets, proposed by Dutch astronomer Jan Oort in 1950. Oort's proposed gigantic halo of comets (to this day, there is only indirect evidence of the cloud) surrounds the Sun, with a radius of perhaps 20,000 to 100,000 astronomical units, far from the center of the solar system. Oort's proposal dealt with the origins of many types of comets. He believed, by determining the accurate orbits of comets, that scientists could prove that the halo's innermost members were the source of long-period comets that eventually made their way into the inner solar system. Perturbed by passing stars or even the undulating periodic ride through the plane of the galaxy, the icy bodies are thought to lose their orbital energy and fall toward the solar system.

The Kuiper disk is much closer to our own neighborhood, and, as many astronomers have suggested, is a better way of getting short-period comets (and thus, maybe the apparent burned-out comets we now call asteroids) into the inner solar system. The idea of the Kuiper disk was proposed in 1951 by Gerard Kuiper, a Dutch–American astronomer at the University of Chicago and University of Texas who had quite an astronomical track record in the 1940s, including finding Uranus' moon Miranda and Neptune's moon Nereid and the discovery of an atmosphere around Saturn's moon Titan.

In a book titled *Astrophysics* (edited by J. Allen Hynek), Kuiper explained his theory.[5] He proposed that the solar system did not stop abruptly at the orbit of the last planet, but continued outward in the form of many smaller objects (comets and planetesimals) orbiting the Sun, between 50 and 100 astronomical units from our star. The reason was that the sparser objects on the initial, outer solar nebula past the orbit of Neptune could not gravitationally form a giant planet. Thus, the leftover material traveled in a flattened disk along the general plane of the planets. Kuiper, and other astronomers since then, further postulated that the small objects may be the fossil remains of the dusty disk from which the planets formed, making them some of the oldest objects in the solar system.

But it was difficult to prove that small objects the size of asteroids and comets surrounded the solar system. The shadowy miniworlds would be faint specks against the black sky and smaller than Pluto, and just think how long it took to find that planet. At the time Kuiper announced his idea, technology had not produced enough light-gathering power, so telescopes were turned elsewhere. The search for the smaller objects possibly roaming the outer rim of the solar system was virtually put on hold for several decades.

By the 1980s, the search began again, one of the catalysts being theorist Julio Fernandez of the University of Montevideo, Uruguay, who postulated that short-period comets may originate in the Kuiper disk, thrown into the solar system by perturbation from larger, unseen planets. Better and more complex computer simulations seemed to confirm the existence of such a disk, with some models even indicating that short-period comets would not exist if there was not a Kuiper belt. Still another impetus was the Infrared Astronomical Satellite (IRAS), which had taken images of dust disks around certain stars, such as Beta Pictoris and Vega. Paul Weissman at the Jet Propulsion Laboratory proposed that the disks were formed by the collisions of comets and asteroids, a condition similar to the theories on how the Kuiper disk formed.

The hunt was on, and it was painstakingly difficult. If there were perhaps hundreds of thousands of the objects beyond about 40 astronomical units, the larger-sized objects would just about reach between 23rd and 24th magnitudes. (Kuiper's original paper estimated that there were 10^{12} objects out there; more recent best estimates range between 10^8 and 10^{10}.) Smaller-sized Kuiper objects would reside in the 28th to 29th magnitude region. Improved search techniques, more sky surveys, and better ground-based, light-gathering telescopes (including CCD arrays and better film for deep sky photographic use) increased the chances of finding objects in the proposed Kuiper disk.

Searches began popping up in several locations. For example, in 1991, a team of scientists from the northern hemisphere (and more recently joined by astronomers in the southern hemisphere), led by astronomer Eleanor Helin of the Jet Propulsion Laboratory, began to search for such objects. The project, called the "Deep Solar System Survey," scanned the sky for objects past Neptune's and Pluto's orbit, with most of the survey sectors outside the plane of the ecliptic, and for several good reasons. First, most of the trans-Neptunian objects have historically been searched for within the plane of the planets, so looking elsewhere would probably prove just as fruitful. Second, Helin's part of the project was tied to the Palomar Planet-Crossing Asteroid Survey at Schmidt telescope at Palomar Observatory in California, a program that looked for objects whose orbits came close to the inner planets. The methods of collecting the data were the same as for asteroids closer to the Earth, but the interpretation of the images was different: The difference between searching for near and deep solar system objects lies in the time lapse between two photos that shows the movement of the objects. Spacing the exposures about 35 minutes apart lets the researcher spot an asteroid near the Earth; but for those farther out past Neptune, exposures have to be separated by a day or longer to see evidence of any movement, and for even more distant objects, by comparing photos separated by two days.

FIRST SIGHTINGS

It was not until August 1992 that the first potential Kuiper disk object, a 23.5 magnitude object called 1992 QB1, was found by David Jewitt of the University of Hawaii and Jane X. Luu of Stanford University. One of the possible tell-tale indications that QB1 was from the Kuiper belt was that the orbit appeared to be nearly circular, which is considered indicative of such objects. The second "miniplanet" (which was the name some scientists started using for the objects) was 1993 FW, found by Jewitt and Luu six months later. Both objects were estimated to be about 200 to 300 kilometers in diameter (to compare, Ceres, the largest asteroid known, is about 1000 kilometers); the objects' magnitudes are between magnitudes 23 and 24, and distances between 40 and 45 astronomical units, respectively.

Another six months of work revealed 1993 RO and 1993 RP, found again by Jewitt and Luu; two more were found by astronomers led by Britain Iwan Williams of Queen Mary College, labeled 1993 SB and 1993 SC. The most recent data show that these four objects have semimajor axes or about 39 astronomical units (5.8

billion kilometers) and orbital periods of close to 246 years (Neptune takes 168.79 years and Pluto takes 247.69 years to orbit the Sun). Some scientists speculate that the quartet of objects may have a 3:2 resonance with Neptune, in which the gravitational attraction of Neptune constantly adjusts the orbits of the bodies into certain specific lanes. This is similar to Pluto's 3:2 ratio with Neptune, and if so, the objects' orbits will never come closer to Neptune than 13 to 15 astronomical units.

Since 1992, the discoveries of these trans-Neptunian objects have increased dramatically. Groundbased telescopes discovered 27 trans-Neptunian objects in the usual heliocentric (Sun-centered) orbits, at a distance of about 40 astronomical units. The objects are all large, averaging about 200 kilometers in diameter and ranging between 50 and 500 kilometers in diameter.

Space telescopes have also added to the numbers of possible Kuiper objects. The best space telescope to make such sweeping forays into the sky is the Hubble Space Telescope. After the craft's launch in 1991 from the space shuttle, an aberration in the telescope optics provided images of the solar system no better, or even worse, than groundbased images. Fortunately, in 1993, another space shuttle mission fixed the Wide Field Planetary Camera, the main unit used to take long-range images of the far reaches of the solar system. With the improved optics, the HST has become an astronomical workhorse, producing clear, often amazing images, and has found additional possible Kuiper objects.

The best results thus far from the Hubble came in early 1995: Anita Cochran from the University of Texas at Austin, and leader of a team of astronomers, found the strongest evidence yet for the Kuiper disk. Astronomers Cochran, Harold Levison from the Southwest Research Institute in Boulder, Colorado, and Martin Duncan of Queen's University in Kingston, Ontario, used 34 deep images taken by the HST's Wide Field Planetary Camera of a single spot along the ecliptic plane in the west-central constellation of Taurus. Eliminating the stars and galaxies in the images, the researchers then combined the images again using two specific offset techniques. One method revealed hundreds of faint objects, mostly noise, near Hubble's limit of 28th magnitude. The other groundbased trans-Neptunian objects mentioned earlier have higher magnitudes, or appear to be brighter, and thus are thought to be larger.

But the researchers also detected objects moving in prograde, low-inclination orbits from the second technique, about 30 cometlike objects that may be indicative of the Kuiper disk, most with diameters of about 20 kilometers, typical of the larger comets. The total now stands at 57 Kuiper disk objects, 27 being at least 100 kilometers in diameter and 30 between 6 and 12 kilometers in diameter. The smaller sizes of the ones most recently found seem to correspond to the size of some well-known short-period comets, including Comets Encke, West, and the famous Halley.[6]

The findings of the Cochran team may be the best reason yet for scientists to point their telescopes toward the ecliptic to search for more members of the Kuiper disk. Already, astronomers are predicting the number of objects in the Kuiper disk: Estimates range from 100 million, to a billion, to 10 billion objects residing between 30 and 50 astronomical units from the Sun. If more objects are found via groundbased or space telescopes, orbital and physical data may definitely prove that the Kuiper disk objects are the reservoir for short-term comets. And stretching it even further, these objects may have the potential to eventually become asteroids in the inner solar system. In fact, some astronomers suggest that the objects of the Kuiper belt should be labeled the *second asteroid belt*.

Scientists believe that early in the formation of the solar system, similar to what occurred around the planets of the inner, and possibly the outer, solar system, the Kuiper disk objects gained mass by sweeping up as much material as possible. But as the Sun went through its "windy" T-Tauri phase (and maybe additional phases that resembled lesser T-Tauri-type stages), it blew away a great deal of dust, gases, and debris from the inner and outer solar system, also wiping out much of the smaller debris from the Kuiper disk. What remains are the larger miniworlds we apparently are finding at the edge of the solar system.

But why did the Kuiper objects form in the first place? Are the "new" objects representative of a once-conjectured "Planet X," now broken apart by whatever tidal forces affected it long ago (perhaps a passing star)? Or are they precursors of many larger, icy bodies, similar to the Pluto–Triton (the largest moon of Neptune) "ice dwarfs" proposed independently by Alan Stern of the Southwest Research Institute and William McKinnon at Washington University? Or could they be similar to the dusty disks currently found swarming around the stars Vega and Beta Pictoris, thought to the remnants from forming solar systems? And of course, there are the questions about the physical structure of the disk itself, such as how far it reaches beyond the orbit of Neptune, how massive it is, and whether it could have enough potential to form planets even at this time.

Right now, finding the objects or determining their origin are not our only concerns. What do we call these strange objects outside the solar system? Described as small, shadowy, dark, and icy, do we refer to them as comets or asteroids? The representatives of the Kuiper disk appear to be a cross between the two: Two bodies lacking cometlike comas, about a half size larger than most comets, and with unusual reddish coloration.

With the addition of these "new" bodies, semantics raises its head: What is the true definition of a comet or an asteroid? As we have noted, it appears to be possible for a comet to become an asteroid by burning out all its volatiles in runs around the Sun. Some asteroids may actually be dormant comets, covered by a thick crust that holds in the volatiles; only with heating from the Sun (or some mechanism we do not understand) do they become what we call comets.

COSMIC QUANDARY

In the middle are those objects that are questionable, crosses between comets and asteroids, sometimes blasting gases, sometimes remaining dormant. One in particular is 2060 Chiron, an apparent amalgam of a comet and an asteroid. In 1977, astronomer Charlie Kowal found the object on a photographic plate in the region of the constellation Aries. The object resembled an asteroid, so it received a new minor planet designation: 1977 UB. After the orbit was calculated, it became 2060 Chiron.

Calling Chiron an asteroid, or a comet, as many had started to label the object, became a dilemma. Asteroids are not usually found in a 51-year orbit that reaches from just inside Saturn's orbit outward, almost to Uranus' orbit. The orbits are usually tighter and inside the orbit of Jupiter. Estimates were also coming in concerning the object's reflectance, and thus its diameter, which ranged between 130 kilometers and about 350 kilometers, depending on the study. More accurate and recent measurements lean toward 250 to 350 kilometers, too small to be a planet, and much too large for the 3- to 10-kilometer diameter of the usual comet. Kowal searched the sky for more such objects in a deep-sky survey concentrating along the ecliptic (reaching down to magnitude 20), and found nothing.

Today's measurements of the object have yielded a wealth of information, including some interesting, albeit puzzling, results. Chiron rotates in 5.92 hours and changes in brightness by about 9 percent at regular intervals, indicating an irregular shape or dark and light spots on its surface. But its regularity ceased in early 1988: David Tholen of the University of Hawaii noted a sudden brightening of the object, along with colleagues William Hartmann, Karen Meech, and Dale Cruikshank, whose measurements confirmed the fact that Chiron's brightness had almost doubled.

Adding to the puzzle, in 1989 Karen Meech, along with Mike Belton of the National Optical Astronomy Observatories, detected a minute coma of ice and dust around the object, one that often swelled to more than 320,000 kilometers in diameter. And in 1990, Bobby Bus, now at Massachusetts Institute of Technology, Edward Bowell of Lowell Observatory, and Mike A'Hern of the University of Maryland detected cyanogen gas in the coma, the first indication of gases around Chiron.

Additional Chiron measurements showed that the object brightened, then dimmed, over weeks or months, by 30 to 50 percent. There were also reports of changes in brightness over the course of a night, although the changes were only a few percent, apparently caused by the amounts of dust, ice, and gases changing in the coma. But perhaps the most amazing puzzle was discovered by a team of astronomers lead by Bobby Bus: Combing through images of Chiron on plates between 1969 and 1972, before the object was discovered, revealed that it was even brighter when it was farther away at 19.5 astronomical units (as of 1996, the object was about 13 astronomical units from the Sun).

What could cause such changes so far away? It cannot be water-ice, as the object is too far from the Sun to sublimate. Carbon dioxide ice would be hard-pressed to sublimate this far out, too. The closest model may lie in the distant solar system: Chiron's reactions may be similar to the surface ices of Neptune's largest moon Triton, seen as dark geysers on images sent back by the Voyager spacecraft. Or maybe Chiron's surface is similar to that of Pluto, possibly made of nitrogen, methane, or carbon monoxide ice. Add the possibility of a thick crust that prevents much of the ice on the surface from becoming active and allows for only occasional vent or fissure eruptions, and it's easy to see why the brightness changes in intensity for short periods of time. In fact, scientists estimate that the amount of material needed to form a coma around Chiron would be about only 0.1 to 1 percent, and such small, sporadic eruptions would be enough to supply the material.

But there is still a major scientific conundrum: What do the scientists label something that is too big to be a comet and too distant to be an asteroid, one that seems to have changes in its cometlike coma at such distances? Again, the Kuiper disk appears as the solution, making 2060 Chiron possibly the first Kuiper object found.

The singularity of Chiron may also reveal a solar system secret. Kowal did not find any more Chiron-type objects in his survey, probably because of planetary gymnastics. The gravitational influences of the giant outer planets—Jupiter, Saturn, Uranus, and Neptune—cause the possible prior population of Chiron-type objects to have short lifetimes in the solar system. In about 1 to 2 percent of the age of the solar system, the planets cause the much smaller objects' orbits to become unstable in just a few million years upon entry into the system. In other words, Chiron's days are no doubt numbered.

NEW DEFINITIONS

The problem of defining these objects will not go away, and eventually they may need a new name. One of the best suggestions so far for the name for these trans-Neptunian objects was made by astronomer Clyde Tombaugh, who suggested "Kuiperoids," from the Kuiper belt, and resembling, more often than not, asteroids.[7]

The findings of the objects in the outer solar system has to be one of the more exciting recent discoveries in planetary science. It gives us a chance to realize that nothing in our system is black and white. There are "new" objects out there to explore and discover, even after we thought we understood the neighborhood.

Will the next "planet" really be called a "planet"? As astronomers find more and more of the small bodies outside the orbit of Neptune, will we find that the solar system's outer rim is filled with another belt filled with small objects? Another asteroid-type belt could be in the offing, and another source for asteroids and comets that enter the inner and outer solar system may be verified.

No one really knows the fate of such trans-Neptunian objects. Most will probably stay in orbit around the Sun in a nearly circular orbit at the distance they are today. Some will become perturbed, perhaps like Chiron, and could eventually wind up as a near-Earth asteroid or comet. In other words, we will probably find out that there are plenty of asteroids to go around, not just from the asteroid belt or from worn-out comets, but another reservoir just ripe for gravitational dipping.

Chapter 9

THE SCARRED AND CRATERED EARTH

*We shall not cease from exploration, and the end
of all our exploring will be to arrive where we
started and know the place for the first time.*
T. S. ELIOT
SNEAKING UP ON EARTH

I live in central upstate New York, a land where, for about 2
million years, the North American Ice Age dominated the
terrain. Climate changes during that time caused huge ice
sheets to advance and retreat about four times, which ground
down layer upon layer of hardened rock, taking away sediment
from hundreds of thousands of years of deposition. As the ice
made its final pass over the landscape, it completed digging the
deep north–south trending troughs that stand as the Finger Lakes,
modified the course of the Susquehanna River, dropped a drumlin
field near Syracuse, and left behind long lines of sediment, such as
a terminal moraine, or a deposit at the end of a glacial ice sheet,
called Long Island. Exposed in midstate were the dull-colored,
395-million-year-old Devonian shales and sandstones, sporad-
ically capped by a hodgepodge of stones and dirt called glacial till.

What was also left was a scraped and barren landscape. And
not many years—just over ten thousand years—have slipped by
since the last of the ice retreated, allowing vegetation to rediscover
the area. But patches of grass or stands of trees often cannot hide
the crushed rock or the long scratch marks from the dragging of
rock and ice across rock. It cannot dissipate the rounded kettle
holes and lakes created by stagnant boulders of melted ice or the
andlayers of rounded pebbles and cobbles laid down when the
meltwaters rolled and tumbled the glacial deposits, all of this as

testimony to one of the most important natural catastrophes that ever affected the region.

Such a huge ripping away of the countryside does not happen everywhere, so it surprised me when, on a flat plain that lies within the San Francisco volcanic field in Arizona, I witnessed a place almost as barren and broken, displaying crumpled, crushed rock. It was another place dominated by natural catastrophe: Meteor Crater near Winslow. But instead of the hundreds of thousands of years in the making, the object that made Meteor Crater took all of about 5 seconds to do its damage.

I visited Meteor Crater (also referred to as the Barringer Meteor Crater) in all its upheaved glory in 1991, while attending the Asteroid, Comet, and Meteor Conference. The meeting, held in Flagstaff, Arizona, brought together some of the top "small bodies" experts in the world, including geologists and impact specialists who study the long-time wanton batterings of the Earth's crust by objects from space.

Holding the meeting near Meteor Crater reminded us all why we were there: Formed by a 300,000-ton asteroid that punched a hole almost a kilometer across in Arizona's Colorado Plateau, Meteor Crater is one of the youngest impact craters in the Earth's crust. Formed about 50,000 years ago when the area was similar to what it is today (although some say a serene pine forest surrounded the area), the impact erased every trace of life for miles around.

Scientists speculate that the impactor was a 50-meter-wide asteroid traveling more than 11 kilometers per second. It entered the atmosphere at about a 45-degree angle and to the southeast, spitting shards of itself as it was pulled into an inevitable collision with the Earth's crust. The impacting body was as bright as the Sun, and it hit with a blinding flash. Immediately after, the blast wave and noise of the collision were far from gentle. The blast created a tall column of dust and debris and turned flat sedimentary beds into crumpled and overturned outcrops, as if someone had carelessly scooped out a piece of the Earth. The noise must have been deafening. A rain of rocky debris fell back to the Earth, while the dust from the impact (rich with tell-tale iridium, an element associated with objects from space) was carried by the prevailing winds from the west.[1]

If a much larger impactor had struck the Earth on the plains of Arizona, the results would have been much more traumatic, affecting the entire planet, as the dust and debris entered the upper atmosphere. As it was, the Meteor Crater strike had mostly a local effect, no doubt vaporizing the surrounding life for miles and miles around. Also in its wake, the impacting body left a boxy-shaped hole, so shaped because of the way the sedimentary rock layers warped around its rim (sandstone and limestone from various local formations, such as the Moenkopi sandstone and the Kaibab limestone).

Nature has been kind to the crater, keeping much of its shape, because the current surrounding desert does not eat into the crater as would a much more humid environment. As for the impacting body itself, scientists believe the majority of the stony, iron-rich asteroid that created the hole vaporized during impact, leaving only small fragments of the original object scattered around the crater.

PROOF OF IMPACTS

It has only been since the early 1900s that we have accepted the idea that asteroids and comets do strike our Earth. Meteor Crater stands as a testimony to the eventual acceptance of impacts, in particular, in work of Daniel Moreau Barringer, a mining engineer from Philadelphia.

I met Mr. and Mrs. J. Paul Barringer (Daniel Moreau Barringer's grandson) while attending the Asteroid, Meteors, and Comets Conference. Both Mr. and Mrs. Barringer were very proud of the work for which the Barringer family is noted. And as Mr. Barringer told me stories of his grandfather (and father, Daniel Moreau Barringer, Jr.), I felt as if I had been listening to the recreation of an historic event: Meteor Crater, not necessarily its discovery, but its subsequent scientific examinations, probably gave the most credence to the idea of catastrophic "impacts from space."

Terrestrial impact cratering, and even cosmic target shooting on the planets and our Moon, was not accepted by the majority of the scientific community before the early 1900s. Even as far back as 1803, after several scientists mentioned the idea that objects could actually fall from our sky, the majority scoffed publicly at such theories. The majority believed that the impact craters were produced from processes of ancient, and now inactive, volcanoes. After all, volcanoes had calderas, the huge circular vents that allow hot magma to release its pressure to the surface. Active volcanoes left large circular holes in the ground, obvious evidence that the craters were formed in the same way. Such ideas endured in the geological world until about 1880, when British astronomer Richard Proctor suggested that the Moon's craters were from the impact of meteoric origins. Even then, the skeptics held on to the volcanism story.

On the more romantic side, Meteor Crater was revered by the Indians, who used the site for religious purposes. Legends from cultures around the region relate stories of a god descending from the heavens, surrounded by fire before coming to rest on the Earth. (It is unlikely that ancestors from the more recent cultures witnessed the event 50,000 years ago, as their ancestors were probably nomadic; and if they did see the event, it would be amazing if the legend held up for thousands of years. But it is possible.) The crater was discovered in the 1870s, as American settlers moved in their westward expansion. First called Coon Butte (or Coon Mountain), the feature was thought to have originated when a volcano erupted millions of years ago. The belief was not at all illogical; after all, the San Francisco peaks, caused by erupting volcanic vents, dotted the landscape not too far from the Meteor Crater site.

Before Barringer arrived on the scene, there were plenty of chances to declare the crater as originating from an impact. The iron chunks found around the crater were collected by prospectors and brought to the attention of Dr. A. E. Foote, a mineralogist and meteorite collector from Philadelphia. Noting that the rocks resembled meteorites, he headed for Coon Butte in 1891, where he collected over a hundred meteorites. After careful analysis, he noted that none of the rocks carried a volcanic signature, but, similar to many scientists of his time, he kept his findings to himself. It would do no good to tarnish his reputation with such a nonacceptable, countermainstream theory. Even the Moon's craters were thought to have originated from volcanism—and what was good for the Moon was good enough for the Earth.

Daniel Moreau Barringer's discovery of the crater came in 1902. His interest in the crater was so great that he acquired the mining rights to the land. Barringer was convinced that the crater was formed by a giant iron meteorite. Forming the Standard Iron Company, he drilled bore holes in search of a meteorite buried under the crater. The meteorite would truly be a treasure, as the fragments he found around the rim contained traces of platinum, nickel, and even small diamonds. After spending about a quarter of a million dollars, and not finding a body that even resembled a mother lode of ore, Barringer was forced to abandon his search because of dwindling funds. He died soon afterward.[2]

But Barringer's Meteor Crater studies were a major catalyst in a paradigm shift, and his evidence finally received a scientific nod in 1955. His contribution was the push toward acceptance of impact cratering as a major physical process on Earth and other planetary bodies. And probably just as important was the eventual realization of what Meteor Crater represents: A small, yet extremely destructive body struck the Earth, leaving a blatant blemish on the planet's surface only a few tens of thousands of years ago—geologically speaking, only yesterday. If an object that size fell on a small town today, it would reek havoc and mayhem, producing an explosion thousands of times worse than those caused by the atomic weapons used at Hiroshima and Nagasaki.

EARTHLY EVIDENCE

As mentioned, Barringer's pronouncement (and others who jumped on the terrestrial impact bandwagon) had an uphill battle in the scientific community. It took until the 1960s for scientists to recognize fully the first impact craters on Earth. Before then many scientists still insisted that most of the circular or boxlike craters were remnants from ancient volcanic explosions, and if impacts were something that occurred on Earth, no doubt they occurred billions of years ago, with the Earth erasing all traces (see figures 1 and 2).

Today, scientists know that Meteor Crater is only one among many impact craters. About 139 impact craters (some claim 120 or up to 200) have been recognized around the world, including Meteor Crater, each with its own size, age, and impact history. Most are less than 200 million years old; some formed quite recently, including the 3353-meter-wide New Quebec Crater in the hard tundra of Quebec, Canada, thought to be *only a few thousand years old*. They are boxy, circular, and even oval in shape; many are merely scars on the Earth, shrunken and difficult to discern because of millions of years of erosion. They are found mostly on the continents, with the potential for many more unknown craters in the oceans (several have been discovered, including a 50-million-year-old, 60-kilometer-wide crater found off the coast of Nova Scotia, and a possible 322-kilometer crater in the Amriante Basin about 483 kilometers northeast of Madagascar).

The impact craters vary in size, ranging from the smaller Sikhote-Alin crater in Siberia, Russia, at 33 meters in diameter, to the Manicouagan Reservoir in Quebec, Canada—a flat, water- and sediment-filled crater measuring close to 70 kilometers in diameter. If there is a bias in the impact craters, it is that they are not usually found in the high latitudes. Remember this before you travel to your next exotic, warm vacation spot: Most of the craters tend to center around the Earth's girth, no doubt a reflection of the orbits of the asteroids, which are usually in line with the orbital plane of the planets. Other reasons include the geologic stability of the regions, allowing the craters to exist even after millions of years, and the fact that more search programs have been conducted in these areas. A table of the major impact craters are listed in Table 1 (see figure 3).

Observing the other bodies in the solar system, it is remarkable that the Earth has remained so seemingly unscathed from the strike of asteroids and comets. But perhaps under all the sediment deposited over the years are hundreds, maybe thousands, more

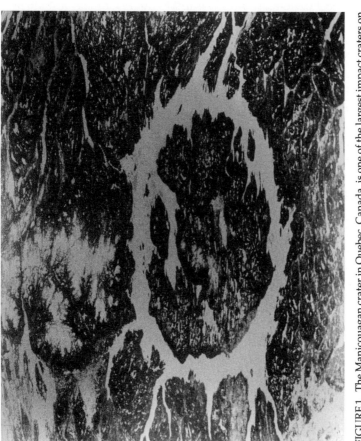

FIGURE 1. The Manicouagan crater in Quebec, Canada, is one of the largest impact craters on the Earth. The 70-kilometer-diameter crater has been eroded by glaciation; scientists believe, based on physical features surrounding the crater, that the crater's rim was originally about 100 kilometers in diameter. (Space Shuttle photo courtesy of NASA)

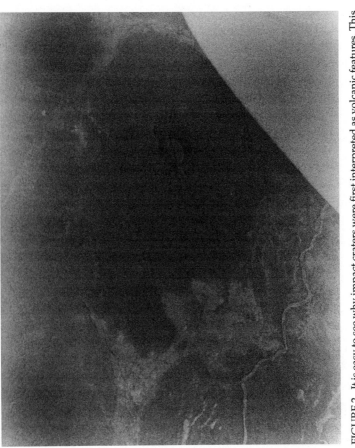

FIGURE 2. It is easy to see why impact craters were first interpreted as volcanic features. This structure of the Gross Brukkaros in Namibia is about 3 kilometers in diameter and about 82 million years old. It was not caused by a space object but a volcanic eruption. (Space Shuttle photo courtesy of NASA)

Table 1. Major Impact Craters on Earth

Location	Crater name	Approximate diameter (meters)
Quebec, Canada	Manicouagan Reservoir	60,960
South Africa	Vredefort	39,624
Germany	Nordlinger Ries	25,146
Saskatchewan, Canada	Deep Bay	13,716
Ghana	Lake Bosumtwi	10,058
Ohio, U.S.A.	Serpent Mound	6400
Tennessee, U.S.A.	Wells Creek	4877
Austria	Kofels	3962
Ontario, Canada	Brent	3658
Ungava, Canada	Chubb	3353
Quebec, Canada	New Quebec	3353
Iraq	Al Umchaimin	3200
Tennessee, U.S.A.	Flynn Creek	3048
Germany	Steinheim	2515
Ontario, Canada	Holleford	2438
Texas, U.S.A.	Sierra Madera	1981
Algeria	Talemzane	1829
W. Sahara Desert	Tenoumer	1829
Arizona, U.S.A.	Meteor Crater	1219

impact craters. According to Australian astronomer Duncan Steel in his book *Rogue Asteroids and Doomsday Comets*, based on a study of four Australian craters smaller than 200 meters in diameter and less than 6000 years old, over the past 3 billion years or so at least 2 million such craters formed on the continent. And the numbers are probably higher: The majority of craters counted were created by iron meteorites, but there are other types of bodies (stony meteorites, comets, etc.) that can form craters, either by contact with the ground or by exploding above the surface.[3]

Finding surface structures formed by an impacting asteroid or exploding comet is not always easy. Some craters are obvious holes in the surface, such as Meteor Crater in Arizona. The majority tend to be helpless victims of the Earth's dynamic nature, torn, bent, fractured, tipped, eroded, and filled with sediment. As ghostly features of their former selves, they are eroded by humid environments and filled in with sediments (such as the Mani-

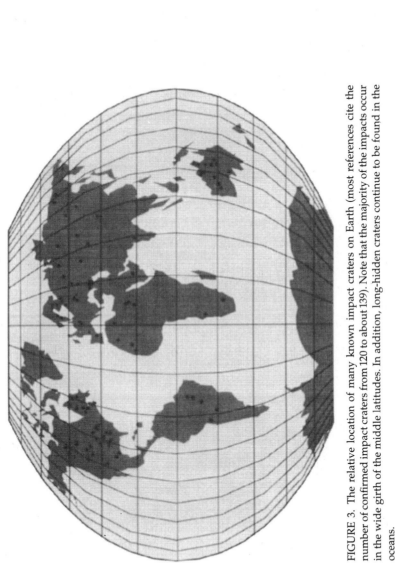

FIGURE 3. The relative location of many known impact craters on Earth (most references cite the number of confirmed impact craters from 120 to about 139). Note that the majority of the impacts occur in the wide girth of the middle latitudes. In addition, long-hidden craters continue to be found in the oceans.

couagan in Quebec), broken or misshapen by the violent movements of the crust, buried in sediment in the oceans, or covered with the thick vegetation of a rainforest jungle.

Because there are also questionable impact craters, those still thought to be ancient volcanic remnants, many scientists use telltale features to identify an impact site. One in particular was uncovered by Robert S. Dietz, an emeritus professor of geology at Arizona State University, who was the first to recognize certain features called shatter cones. As a shock wave races out in front of the impacting body, the waves shatter the rock in a distinctive type of fracture. The deformation resemble fans within the rock—a fossil of the shock wave generated the instant the impact was irreversible. If the rocks were put back in their original position, the points of the shatter cones would point inward toward ground zero in the center of the impact basin.[4]

There are other smaller features: Rocks around an impact site usually exhibit metamorphism from shock pressures that exceed millions of atmospheres and temperatures above 2000 degrees Celsius. In particular, such high pressures affect quartz deposits, the high pressures creating a mineral called stishovite. Some rocks also register microscopic changes in certain minerals around the impact site. For example, pressure lamination (seen as tiny, thin lines) in quartz minerals are most often associated with the blast from an colliding asteroid. Fused silica bombs, called impactites, are also indicative of impact craters, small, black masses composed mostly of glass (with minute iron fragments throughout) that are formed as the impact melts surrounding silicate rocks.

Larger features also can be indicative of an impact, including a collection of jumbled rock produced by an impactor striking in the oceans. A huge wave in the oceans is usually called a tsunami (Japanese for "harbor wave"; it is often incorrectly referred to as a tidal wave, but unlike an actual tidal wave, it is not caused by the pull of the Moon or Sun on the Earth's oceans). A true tsunami is caused by the sudden movement of the ocean floor during an earthquake, creating a displacement of the water, and thus a large sea wave. In the case of the asteroid impact, the result would be similar, but much more drastic. Perhaps the terms *impactnami* or *imnami* would be more appropriate for the waves produced by impacting asteroids!

There are good reasons to distinguish the usual seismic-generated tsunami from a wave caused by an impact. The impactor's wave and the damage it could cause would be far greater than that from the usual tsunami. For example, seismic profiling (reading echoes from sound bouncing off various densities of underground rock) by several researchers at the United States Geological Survey in Woods Hole, Massachusetts, has led to the possibility of a new find: a meteorite crater about 85 kilometers in diameter at the mouth of what is now the Chesapeake Bay. The scientists believe that the site is a buried impact crater about 35 million years old and that the impact, which occurred 100 kilometers offshore, created a huge sea wave as the impactor hit the water. The evidence includes the jumbled sediments reaching all around Virginia and the Delmarva peninsula, material carried by the huge wave. If an asteroid struck in the same place today, the damage and destruction would be unimaginable.[5]

Even with all the physical evidence gathered from impact craters, there are still questionable sites. One of the most debated sites is found in Sudbury, Ontario, Canada, a mineral-laden 64-kilometer-long and 32-kilometer-wide structure that many scientists still insist is a natural terrestrial deposit. The majority of the Earth's nickel deposits are associated with ultrabasic rocks pushed up from the Earth's mantle, a layer of the planet just below the crust. And in this case, the region is an ultraproducer: The Sudbury Basin yields about 75 percent of the West's nickel and about 20 percent of all Canada's mineral wealth.

What is the best evidence that Sudbury is from an asteroid impact? First, the area rocks are not associated with any surrounding mantle rocks. Second, there are small patches of shock-melted rock called pseudotachylite found as far away as 48 kilometers from the impact, formed as the impact's shock waves criss-crossed and dissipated. And finally, even though the impact site has gone through about 2 billion years of deformation, the rumpled rock contains shatter cones around the alleged impact site. With all the evidence, most scientists agree that the metals taken from the prolific mine are from space (another similar nickel mine in north-west Siberia, called Noril'sk, is thought to also be from an asteroid impact), but not everyone wants to concur.[6]

FAKING IMPACTS

Determining how the impacting bodies created the craters has long been a matter of conjecture. More recently, one of the best ways scientists have had to study the way impact craters make an impression on Earth is essentially to play with explosives, especially offshoot techniques of nuclear testing. By bombarding small-scale models of the Earth's surface with small, proportional projectiles at relatively similar speeds, scientists have determined many of the dynamics and results of a variety of asteroid strikes. But our own feeble attempts to imitate an asteroid impact is nothing (thankfully!) compared to what nature can offer us at the actual site.

Test explosions in the field have given us some idea of the impact effects, and it is not a pretty sight. Blasting a 100-kilometer-wide crater would involve the excavation of several times 10^{15} tons of rock, which would be the energy equivalent of more than 10 trillion tons of TNT. If only about 1 percent of the ejected dust and debris were to reach the upper atmosphere, the Earth would be covered with a thin layer of dust several centimeters thick. The aftermath would also mean months of dust in the atmosphere, so thick that the sunlight would be dimmed, the entire Earth becoming the land of the midnight sun.

Other teams of scientists are also using techniques from nuclear testing to understand impact sites. For example, Randall T. Cygan and Mark Boslough at Sandia National Laboratories, along with R. James Kirkpatrick of the University of Illinois, draw on their work for the Defense Nuclear Agency, using nuclear magnetic resonance (NMR) spectroscopy to analyze minerals shocked by extreme pressures. The technique has enough sensitivity to determine how heat and pressure changed the structure of the minerals within the samples. Such minute distortions at the atomic level of shocked minerals are not possible to find through traditional methods, such as x-ray diffraction and electron microscopy that requires each mineral to be individually analyzed. The intent of the study is to provide a calibration for determining shock pressures in battered minerals, thus creating a kind of impact "shock barometer." Using experimentally shocked quartz as their base, they applied the technique to Coconino sandstone from Meteor Crater, Arizona, finding stishovite and coesite and even a new high-pressure phase in a moderately shocked sample, a dense hydrous amorphous silica, and from the Cretaceous–Tertiary boundary from Raton Basin in New Mexico. They are also beginning to look at other NMR active nuclei found in carbonate rock (common to Chicxulub target rock) to see how sensitive the NMR signals would be to shock pressures of impacts.[7]

BURIED DEEP

Probably the most frustrating aspect of impact crater searches is knowing that most of the older craters are buried deep within the layers of the Earth, never to be found. Some have been found purely by accident, usually as the result of the area being explored for gas or oil. For example, the now-famous buried Chicxulub crater located on the Yucatan Peninsula in Mexico, which will be discussed in detail later, has been tied by many scientists to the massive extinctions that took place at the end of the Cretaceous period, 65 million years ago. The circular feature was found because of an intense study of the area in the search for oil, and even though the data were taken at the site in the 1960s, it took many more years until the circular feature was finally discovered to be an impact crater.

The longtime techniques of gravity measurements and seismic profiling to find oil and gas reserves, in conjunction with other methods such as deep drilling, which reveals layers of rock deep below the surface, have led to several possible underground craters. Using these various methods, several alleged buried, somewhat circular features have been found and analyzed for their potential as impact craters. Three of the better-known craters explored in this way are the Manson and Ames structures and the already mentioned Chicxulub crater.

The Manson structure, found in Iowa, is estimated to be about 35 kilometers in diameter, too small to have wreaked global havoc on an unsuspecting world. But until recently, scientists believed that the small underground crater had a great deal to do with the extinctions at the Cretaceous–Tertiary boundary, especially as an accomplice to the much larger Chicxulub crater in Mexico.

In 1992, Eugene M. Shoemaker and Glen A. Izett of the U. S. Geological Survey determined that certain 65-million-year-old deposits near Trinidad, Colorado, showed two separate extraterrestrial impacts—a lower region several centimeters thick and an upper one only a few millimeters thick. Originally, the two layers were thought to be from the same strike: The lower layer representing the initial ejecta from the impact, and the upper, thinner layer representing the light rain of dust after the impact.[8] Shoemaker and Izett believed that the layers formed by two separate impacts occurring a few years apart. Shoemaker showed several areas in a rock layer that had established plant roots, something that could not have happened if the layers represented only one impact.

The search was on for not only one large impactor, but also another small one. Chicxulub already seemed like the logical choice for the first impact at that time; the smaller alleged impact centered on the Manson structure. The proposed diameter of the circular feature would be large enough to cause the second layer to form; all it would take is drilling the area to seek out the evidence.

But by early 1992, several studies effectively removed any chance that Manson played a part in the Cretaceous–Tertiary boundary extinctions. Drilling done in 1991 revealed that the rocks near the structure that melted and recrystallized had normal magnetic polarity, whereas the Cretaceous–Tertiary boundary occurred during a period of reversed polarity, in which the north and south poles switched. The Earth's polarity, for reasons yet unknown, reverses every few hundred thousand years, usually taking about 2000 to 20,000 years to complete (the most recent one took place about 730,000 years ago). In addition, little is known as to the effects of such events, especially on lifeforms. Currently, lines of magnetic force flow downward at the north magnetic pole and upward at the south magnetic pole; it is close to opposite at times of polar reversal (the complexity of the reversals makes the new pole locations fall within certain regions, which are not exactly opposite the preceding magnetic pole). The reason scientists know so much about the reversals is because of rocks: When volcanic magma solidifies, its small iron crystals behave like compasses and lock in any magnetic dip and direction of the north magnetic pole at that time.

Because the polarity of the Manson rocks did not correspond with the Cretaceous–Tertiary polarity, the scientists reasoned that the feature is either younger or older than 65 million years, but it cannot be that exact age. Thus, the Manson structure has been disqualified as a Cretaceous–Tertiary boundary damaging impact event.[9]

Another study actually dated the rocks: Using the ratio of argon isotopes 39 and 40 dating technique, the rocks were found to be 73.8 ± 0.3 million years old, too old to have been effective in the Cretaceous–Tertiary boundary event. Not that the impact was benign. Some scientists believe, based on the shallow water regime that existed in the region at that time, that the Manson strike could have caused a relatively destructive tsunamilike wave. But so far, no such material has been discovered in the surrounding region. So the search continues for an impactor that joined Chicxulub.

There are other buried impact structures around the world, many of them not associated with an extinction, but just as valuable in determining Earth's past impact history. The Ames structure, located on the Anadarko basin shelf in northwestern Oklahoma, is a prolific source of oil and gas from more than 50 producing wells. Buried beneath 2743 meters of sediments, the circular structure was found in 1991. It is thought to have formed during the early Ordovician, making it about 500 million years old. But not everyone agrees; some scientists argue that the structure is merely a volcanic crater or a weak feature that collapsed, forming what appears to be a crater.

There are other questionable craterlike features buried in the depths of the Earth: the Marquez Dome, Wells Creek, Haswell Hole, Calvin, Big Basin craters, Kentland Dome, Red Creek, and even a postulated 3000-year-old crater in central Nebraska. And the list continues to grow, as do studies to prove or disprove the cosmic origins of these craters.

NOW YOU DON'T SEE IT

Not all impacts leave large holes in the Earth. Some end their existence high in the atmosphere, unable to shrug off the effects of plunging through the atmosphere. More recently, because of frequent air flights and additional areas set up for seismic data collection, we have found evidence of such visitors from space. For example, on September 22, 1979, a bright flash was detected by the Navy's surveillance satellite Vela near Prince Edward Island off South Africa. A nuclear weapon was ruled out. A natural event—an exploding comet or asteroid—was the best explanation.

Not all the material is small, either. Astronomer Eugene Shoemaker noted that we get an object with about 20 kilotons of TNT energy entering the atmosphere every year, but it never reaches the ground. In comparison, the Chicxulub object, thought to have helped in the Cretaceous–Tertiary extinctions, struck with the force of a 100-million-megaton bomb, an event that probably occurs only once every 100 million years or so.

Apparently, military satellites have been watching such meteoroids strike the Earth's atmosphere for almost two decades. From 1975 to about 1992, infrared scanner data from military satellites were recorded by the U.S. Department of Defense satellites. The information was classified until recently, and released data has shown that 136 atmospheric explosions have been detected, with yields of 1 kiloton or more in that short 17-year period. The actual number may be more roughly 10 times more, but the satellites were programmed to watch for unnatural events, such as nuclear detonations. There were also gaps in the satellites' coverage, and because the meteoric flashes occur only for a few seconds, if an array did not cover the area, the event was missed.

Such explosions are mostly seen as infrared events, with few of the exploding projectiles emitting enough visible light to be seen by an observer on Earth. But those that can be seen may explain many of the strange reports of sky phenomena over the past two centuries. For example, on April 15, 1978, a military satellite watched a huge fireball (yield about 5 kilotons) in the daytime over Indonesia that, for one second only, would have rivaled a burst of sunlight for anyone watching from the ground below. Another report on August 3, 1963, noted a huge airburst equal to 500 kilotons between South Africa and Antarctica. The burst was picked up by a worldwide network of acoustic detectors; it is thought to have been caused by a small asteroid about 20 meters in diameter. A burst as bright as the Sun some 30 kilometers in altitude occurred on October 1, 1990, luckily occurring over the western Pacific Ocean, and not over the fighting in the Gulf War occurring at that time.[10]

Another suspicious occurrence occurred on April 9, 1984. As we noted earlier, the pilot of a Japanese cargo plane reported seeing an asteroid explosion about 644 kilometers east of Tokyo, Japan. The blast formed a mushroom cloud, rising from about 4267 to 18,288 meters in only 2 minutes.[11]

LESSER IMPRESSIONS

Smaller bodies also can create quite an impression on the Earth: A 1 kilometer asteroid, according to some estimates, collides with the Earth every 1 million years. An object this size would produce an explosive force comparable to several hydrogen bombs and would leave a crater 13 kilometers across.

Though not all such bodies form an impact crater, they can still impact the surface of our planet. For instance, one of the most famous stories of such a body took place in the northern Siberian region of Tunguska, on June 30, 1908. The early explanations for the massive destruction found around the site centered on electrical disturbances caused by solar outbursts in the northern atmosphere; as more data was collected on the site, the theories ranged from the crashing of a UFO to a mini-black hole destroying an area about the size of Washington, DC.

Eventually, the evidence pointed to an extraterrestrial visitor, a comet with such low density that it would rapidly decelerate, essentially stop in the atmosphere, and then explode. But subsequent computer models showed that it would be difficult to explain the explosive episode with such a comet. After all, knowing a little more about a comet's nucleus, researchers determined that the comet would explode too high in the atmosphere to account for the Tunguska event. They turned to an asteroid, a stony asteroid only 30 meters in diameter (others claim diameters closer to 60 meters; and even the size of a city office building) moving at roughly 15 kilometers per second. They determined that as it hit the Earth's atmosphere, it would disintegrate at roughly the same height as the Tunguska object exploded. In order to respond like a Tunguska event, such an asteroid would have to be extremely strong. A carbonaceous asteroid would explode too high under such conditions; an iron asteroid would explode too low, if at all (in order for an iron asteroid to react like a Tunguska event, it would have to be traveling unusually fast).

In 1994, two scientists found the evidence they needed: Solid particles from an asteroid were found imbedded in tree resin from conifers at the site, much like prehistoric insects trapped in amber. The asteroid fragments they found were of the stony type.

Now scientists lean toward the following story: The stony asteroid vaporized as the radiated energy from the shock and pressure in the asteroid's shock wave reached an explosive threshold. Based on airwaves recorded on meteorological barographs in England, the object exploded with the energy of a 10- to 12-megaton airburst, although estimates closer to 20 megatons are often suggested based on physical evidence near the blast site. (The blast has also been described as exploding with 2000 times the force of the nuclear blast that devastated Hiroshima, Japan, but a better and more accurate estimate is a few Hiroshima bombs). The asteroid appears to have entered the atmosphere from the southeast at about a 30-degree angle. The blast occurred about 5 to 10 kilometers above the area, creating a scene of immense destruction: The shock wave knocked down trees within 2150 square kilometers, the trees stripped of branches and leaves, all pointing in the direction away from the blast, while an area half that size was incinerated.

There were no fatalities, but according to astronomer Roy A. Gallant of the University of Southern Maine's Southworth Planetarium, there were plenty of witnesses. The asteroid struck in the morning, reportedly leaving an 800-kilometer trail behind. Most of the eyewitness accounts were from nomads who had camps near the area, some even within the blast site area. Within the radius of the blast, a storage hut of Stepan Dzhenkoul was burned; the 700 reindeer in Vasiliy Dzhenkoul's nomad camp were burned, as were the dogs, all stores, and all the tepees (fortunately for Vasiliy, he was tending another of his herds when the asteroid hit). Other incidences—including the leveling of all teepees in the area and several people falling unconscious or knocked off their feet— were scattered within and just beyond the blast site.[12]

One interesting aside: In July, 1993, a 5-year drilling project through Greenland's ice sheet struck bedrock, producing the longest ice core from the northern hemisphere. A year later, numerous results came out of the Greenland Ice Sheet Project II, based on the finely detailed layer upon layer of ice records, including a possible link with the Tunguska object. Robert Sherrell of Rutgers University, Edward Boyle of the Massachusetts Institute of Technology, and Robert Rocchia of the Centre des Faibles Radioactivités, France, noticed 4- to 20-fold jumps in iridium concentration in the Greenland ice cap that correlated with the time of about 1908. Because excessive amounts of iridium are associated with an impacting body, the scientists are trying to confirm that the iridium spikes are truly from impacts, and if so, this is the first demonstration of a meteoritic impact recorded in ice. Additional ice core checks may reveal more such iridium spikes, but there is cautious optimism: The iridium signature will have to be confirmed first, though, to isolate it from that of a volcanic eruption. To understand the problem, the researchers also found an 18-fold iridium enrichment in the ice that corresponded with a sharp increase in sulfate produced by the 1783 volcanic eruption at Lakigagar, Iceland.[13]

Although Tunguska has been written about for close to a century, mainly because of its visible and extensive effects, there have been other close calls, including another strike in Siberia (Siberia is one big expanse of land, a perfectly huge target for striking bodies): An iron projectile exploded over the Sikhote-Alin region in February, 1947, creating more than a hundred small craters, measuring from 1 to 14 meters in diameter. And not all of the close encounters occurred so long ago. There is the large meteorite that crossed over the Grand Tetons of Montana in 1972, captured on film as it passed overhead, and the 5-meter crater created by a 1-meter iron meteorite in 1990, near the town of Sterlitamak in eastern Russia.

GLANCING BLOW

A skipping space object also left a fleeting glimpse of itself in northcentral Argentina. Ruben Lianza, an Argentinean pilot and amateur astronomer, discovered the group of curious depressions not far from the provincial city of Rio Cuarto, Argentina. His aerial photos of the strange gouges revealed some oddities not common with most natural phenomena: The long, elliptical swaths followed the same direction (the recent, geologically speaking, rips in the surface resemble huge glacial striations, although the area had no recent glaciation), and they diminished in sized to the south. The craters seemed to mimic a pattern similar to gouges caused by glancing, high-speed impacts in laboratory experiments.

At first, such claims were doubted, and it took 2 years for confirmation. Finally, researchers agreed with Lianza's find: A family of impact craters (actually gouges) numbering about 10 in all, do exist. Besides the slices in the Earth, several other pieces of evidence have also been found: Glassy blobs of fused dirt called impactites were found in the area, containing deformed grains of quartz, evidence that a high-energy impact had occurred in the area. Even a chondritic meteorite was found at the site early in its exploration (although it is possible this fell at some other time).

To determine how the elliptical gouges could form, scientists had only to turn to the scarred surfaces of other members of the solar system, and then compare the shapes of the craters to those formed in the laboratory. For example, such glancing blows to the surface occurred at the Moon's crater Messier, measuring only 6 by 14 kilometers, and were similar to craters formed experimentally when an object struck at a 5-degree angle. On a larger scale, but just as strangely shaped, is the 350 by 500 kilometer Crisium basin. Oblong craters on Venus and Mars also exhibit the telltale low-angle collisions of an asteroid. Venus' dense atmosphere probably contributes to the oblique strikes; on Mars, some scientists believe that many of the oblong craters were produced by glancing blows from ancient satellites.

Based on comparisons between laboratory and field studies, scientists worked out how this family of craters formed: Assuming that the velocity was about 23 kilometers per second (typical for a near-Earth asteroid), the scars in Argentina, some 30 kilometers long and 2 kilometers wide, were caused when a 150-meter-diameter near-Earth asteroid struck from the northwest, approaching no more than 15 degrees from horizontal. The asteroid probably released energy equivalent to a 350-megaton bomb, or 30 times that of the Tunguska event and 10 times that of Meteor Crater's energy, as it skipped over the countryside. The first crater to form was the largest (the "Northern basin"); the smaller basins beyond formed as pieces from the original strike sliced open the Earth.

If the strike had been more typically vertical, the results would have been more catastrophic. As it was, the consequences of this strike were probably locally confined, with firestorms spreading out and incinerating life in the area of the impact and some debris entering the upper atmosphere. The age of the strike is still debated, but some scientists estimate it may have occurred after the end of the most recent Ice Age, about 10,000 years ago.[14]

Perhaps the greatest evidence of our luck (after all, it could have struck vertically) is that life was not eradicated on a global scale. The gradual buildup of a worldwide civilization continued to spread after all these visits, and so far we have managed to miss the effects of an asteroidal potshot.

INDIRECT EVIDENCE

I met Virgil Barnes when he worked at the Bureau of Economic Geology at the University of Texas at Austin. Retired from the UT faculty in 1977 (he was born in 1903), he was still continuing his research at the BEG, including work on his favorite subject, tektites. Barnes told me that by chance he became intrigued with the small rocks in 1936 when a geologist working on a bureau project in Grimes County, east Texas, found some unusual rocks he thought were obsidian. But the closest igneous outcrop was in west Texas, and it did not look like any obsidian Barnes had ever seen. Digging deeper to find out more about the strange black rocks, he discovered they were tektites (at that time thought to be meteorites). He also suggested their probable origins: the results of impacting asteroids and comets striking the Earth.

Tektites are small glassy chunks that range from walnut- to grapefruit-size and are button to teardrop in shape; smaller renditions of tektite are called microtektites, which are no larger than a head of a pin, usually associated with sediment found in the oceans. The rocks range in color depending on the location of discovery and are often mistaken for slag or obsidian (the black to dark green glass that forms from volcanic magma). If you look closely at most tektites, they show striations or pock-marks, evidence that the stones were dramatically heated by friction as they entered the atmosphere, bubbled, and then were shaped as they fell to the surface.

The first person thought to have described these strange stones was Liu Sin, a Chinese scientist who lived about 950 A.D. They were collected around the Luichow Peninsula by the villagers after heavy rainstorms. Amazingly, they called the stones lei-gong-mo, "inkpots of the Thunder God," or lei-gong-shih, "the stool of the Thunder God," most likely in reference to the rocks' origins after a rainstorm.

Researchers have only found the tektites in certain regions around the world, called *strewn fields*, each with their own name for the tektites found. Estimated ages of some types of tektites are highly debated, and the most commonly agreed-upon ages are listed here: The yellow *Libyan Desert glass* in Egypt, mostly black *Ivory Coast tektites* along the Ivory Coast; the mostly black North American tektites (estimates range from about 34 to 45 million years old; they are thought to be the oldest tektites on Earth) are found from Texas (bediasites, after a local Indian group) to Canada, and in Georgia (one lone tektite was reportedly found at Martha's Vineyard, Massachusetts). The mostly dark green moldavites around the Czech Republic and Slovakia are about 15 to 20 million years old. The great Austral-Asian area strewn field, which covers almost one-tenth of the Earth's surface, is subdivided into the mostly black australites in Australia (the youngest tektites, about 0.7 million years old); the mostly black javanites in Indonesia; the mostly black indochinites in Thailand, Cambodia, southern Vietnam, and Malaysia; and the mostly black philippinites in the Philippines. There is also a local impact glass from Tasmania, about the same age as the australites, called Darwin glass.[15] (Figure 4)

Speculation on the origin of the rocks have ranged from pieces of a destroyed planet and ancient attempts to manufacture glass, to fulgurites formed by lightning strikes in sand. Today, most scientists heartily agree that the small rocks within each strewn field came from impacting bodies striking the Earth. Apparently, the material from the impacts were thrown to great heights. As the material fell back to the Earth's surface, it was melted and deformed by the frictional heat of the atmosphere, creating the glassy tektites within a specific strewn field, along with a spray of microtektites. There are also those who disagree: For example, astronomer John O'Keefe, at the Goddard Space Flight Center, once proposed that tektites came from the Moon when a large object struck our satellite, sending tons of debris into space and toward the Earth.[16]

Just where those impacts occurred on the Earth has long been a bone of contention. Although there has been a great deal of speculation, no real surface or buried craters have been found to substantiate their origins. According to Barnes, the Ries Kessel formation in southern Germany has been identified as an impact crater, and several scientists have proposed that it is the site from which the moldavites originate. The Ashanti crater in Ghana may also be large enough to have produced the Ivory Coast tektites.

Discovering what caused the formation of tektites may be difficult, as thousands to millions of years of erosion could have erased any evidence. And if the impacts occurred in the oceans, finding them may be impossible if they are already covered with millions of years of ocean sediment. To find the origin of the oldest tektites in Texas may be too much to ask for; after all, scientists would probably have to search the floor of the Gulf of Mexico or the Gulf coastal plain under a mile or so of sedimentary deposits. In addition, the philippinites and the indochinites may also be a lost cause, the impact sites possibly being buried under volcanic lava; according to some scientists, such an impact could initiate the volcanic event that would hide the impact scar.

There is also a problem with associating an impact event with the australites. These tektites are extremely young, on the order of about 10,000 years old. In order to produce such a strewn field, the event would have had to be of sufficient magnitude to blow out a segment of the atmosphere, along with the shower of tektites that would remold upon reentry. The impact site would have to be gigantic, probably just over 160 kilometers in diameter, and such a crater would also be very obvious. Of course, there is another suggestion: According to scientist John D. Wasson, in a paper on the australites in particular, it is possible that the tektites formed as southeast Asia was peppered with many small cosmic projectiles. Instead of looking for a 100-kilometer crater, perhaps there were many, measuring only about 1 kilometer in diameter. If so, the craters may have already been consumed by erosion or sedimentation.

LEADING LEGENDS

More recently, because we know that many of the asteroidal encounters have occurred within the lifetime of *Homo sapiens*, the search for previous impacting bodies has turned to cultures that may have endured such a strike. And a scientist may have found

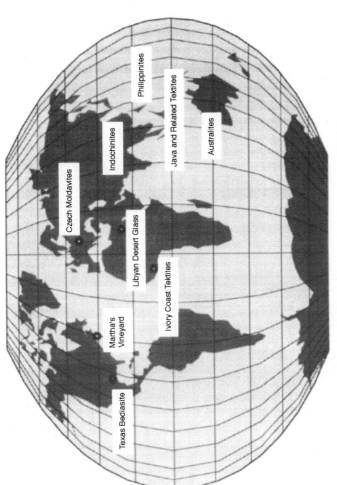

FIGURE 4. Tektites are glassy chunks of rock found only within certain regions, called strewn fields, on the Earth. Certain craters are thought to have created certain strewn fields (for example, the Ashanti crater in Africa producing the Ivory Coast tektites and the Ries Kessel craters in Germany producing the Czech moldavites), but these claims have not been verified. The Martha's Vineyard tektite is the only one found as a single fragment. The oldest tektites are the bediasites; the youngest are the australites. The tektites on this map without a dot means that the strewn field encompasses a large area around the location markers. (Location sites courtesy of Virgil Barnes)

Texas Bediasite

Martha's Vineyard

Ivory Coast Tektites

Libyan Desert Glass

Czech Moldavites

Indochinites

Java and Related Tektites

Philippinites

Australites

such evidence: Duncan Steel, a research astronomer at the Anglo-Australian Observatory and research fellow at the University of Adelaide, Australia, has a proposal based not only on physical evidence, but also, in an unusual twist to find scientific evidence for an asteroid strike, on legends passed down from New Zealand's Maori, an ancient Polynesian race on the lower island. Steel, along with New Zealand physician Peter Snow, believe they have found evidence of a comet or asteroid vaporization that occurred about 800 years ago, similar to the Tunguska event in Russia.

According to Steel, the ovoid crater called Landslip Crater or Landslip Hill in Tapanui was always interpreted as a landslide, but without the usual indications of the land that goes along with a slide. Steel says it seems to be more like a excavation in loose terrain. And unlike a normal impact crater with evidence of shock metamorphism, such a Tunguska-type event shock wave would excavate a different type of crater. Other indications are there, too, including tree falls pointing radially away from the Tapanui site and evidence of flash forest fires dating to about 800 years ago.

What do the Maori have to say about the event? Legends in Maori songs and poetry refer to devastating fires from space, raging winds, and the upheaval of the Earth. There are records of Maos, a now-extinct bird "felled by strange fire." The name Tapanui apparently means "the big explosion," or "the big devastating blow," and even towns around the area have names they relate to fires from the sky.

Could such legends become part of the scientific evidence needed to determine such explosive events? Steel seems to think so, especially since the impact would leave an impression not only on the Earth's crust but also definitely in the minds and legends of ancient cultures. In fact, Steel believes there may be indications of other such events in Brazil and Australian aboriginal records. According to Steel, in mythology, people abandon certain areas frequently, and there are many evacuations that occur around the same time. Scientists usually say climate changes are responsible; but reading the records, many groups seemed to leave at around the same epoch. In fact, he believes there were many impacts around 800 years ago in areas that were disconnected.

The correlation is also being tied to the cycle time of the Taurid Complex, the supposed debris produced by the breakup of a giant comet between 10,000 and 20,000 years ago. The theoretical complex may produce recurrent meteor showers, comets, and asteroids that intersect our planet four times every few thousand years, with each period lasting a century or so. So far, one such period strongly matches the time eight centuries ago when the Maori peoples saw their "strange fires" from space.[17]

All types of evidence keep popping up all over our planet, letting us know that we have not escaped (and never will escape) the attacks from impacting space objects. We have not yet found all the answers to these unpredictable visitors, but we do know our latest impact crater will not be our last.

C h a p t e r   1 0

FIGHTING EXTINCTIONS

*The force of gravity on their surfaces must be
very small. A man placed on one of them would
spring with ease 60 feet high, and sustain no
greater shock in his descent than he does on the
Earth from leaping a yard. On such planets
giants may exist; and those enormous animals
which here require the buoyant power of water to
counteract their weight, may there inhabit the land.*
J. NORMAN LOCKYER (1883)
ELEMENTS OF ASTRONOMY

STRIKES THAT ELIMINATE

We can describe the impact craters left on the Earth by asteroids and comets all we want, but what were their true reverberations on the Earth? Certainly smaller strikes would have had a localized effect on organisms on the Earth. But what about the larger strikes over geologic time? What repercussions did they have on the Earth as a whole?

On the Moon, impactors made (and continue to make) holes in the surface, but really have no effect on the overall planetary body. There is no atmosphere in which to throw the ejected material, and near-Earth asteroids that come close to the Moon are not large enough to break up or even fragment the satellite. Additionally, the Moon is thought to have a much thicker crust and mantle than the Earth because of a lower internal temperature, in other words, it is solid enough to absorb strikes from larger objects.

The only effect of a striking Moon-crossing asteroid on the

Moon might have been a moonquake. The way we know this is simple: Each Apollo mission to the Moon left behind an ALSEP, an instrument complete with a seismometer or "moonquake recorder." The three distinct types of moonquakes recorded were caused by the impact of meteorites, by artificial means (the Lunar Module slamming into the Moon after the astronauts rejoined the Command Module), and by movements in the Moon's interior.

Take a planetary body with more to offer, and an even greater amount to lose; take a planet that is teeming with life, plants, animals, insects, fungus, bacteria, viruses, and everything in between. This was Earth during many of its geologic time periods. On occasion (and there were many), something radical occurrs on Earth, eradicating much of life on the planet for no obvious reason. These times, called extinctions, are due to some natural catastrophe, wiping out species that could not get out of the way or adapt to the changes after the catastrophe occurred.

When paleontologists talk about extinctions, they are not talking about our current loss of species such as the dodo bird or the passenger pigeon. Most of the time, they are pointing to extensive extinctions, times when 50 percent or more of the species died out in the oceans and/or on land. In addition, when paleontologists discuss extinctions, they also disagree on the speed of an extinction. In geologic terms, many of the species disappeared rapidly; other scientists contend that the fossil record is just incomplete and that most of the extinctions took place gradually. No one really knows which is correct, but they agree that the extinctions definitely occurred.

For example, about 245 million years ago, between the Permian and Triassic periods on the geologic time scale, about 96 percent of all species on Earth became extinct (although more conservative estimates say up to 80 percent became extinct). By this time, Earth had experienced about 3.75 billion years of evolution: Blue-green algae began to photosynthesize and liberate oxygen in the oceans by 3 billion years ago, and life-forms in the oceans grew almost exponentially between the Precambrian and the Cambrian, 600 million years ago. Sea pens grew on the ocean floor, and jelly fish, trilobites, and jawless fishes filled the oceans. Land plants began to take hold during the Silurian (about 400 million years ago), as such plants as Cooksonia found a new, drier home.

According to the fossil record, during these millions of years, there were small extinctions, such as those in the oceans about 505 (between the Cambrian and Ordovician) and 438 million years ago (between the Silurian and Ordovician). There was an even larger one, which included the extinction of organisms on land and in the sea, at 360 million years ago, between the Devonian and Carboniferous. Even closer to our own time is the sprinkling of smaller extinction patterns, including those of the Eocene and middle Miocene, 35 and 11 million years ago, respectively.

But none compared to the Permian extinction. A 96 percent figure means that 96 out of 100 species vanished from the Earth. It was the end, not only of the Permian period, but also in terms of the geologic time scale, a change of an era, from the Paleozoic (loosely translated to "ancient" time) to the Mesozoic (loosely translated to "middle" time). The balanced, fairly stable major shallow marine ecosystems collapsed, causing the mass extinction of most marine flora and fauna. Even land organisms became extinct, although most were not as affected, including many species of an animal kingdom known for its profusion in the Mesozoic— the reptiles.

PERMIAN EXTINCTIONS

Just how did these extinctions occur? At this point, I must point out that extinction is no different from any other scientific quandary—it comes with a bundle of theories. Scientists have long debated the demise of species, and the arguments still continue to today. Let's take the Permian period again (also called the late Paleozoic era), which paleontologist Stephen Jay Gould has described as the "granddaddy of all extinctions"—a time far more deadly in terms of species elimination than the events that killed off the dinosaurs 65 million years ago. Was it a giant asteroid, or many asteroids, that caused the heavy loss of life during this time, or was it something else?

It is not certain that impacts caused the life-forms' demise, and there are many good arguments to the contrary. During the Permian, all the world's continents, which looked remarkably different from today's continents, were sutured together in a super-continent called Pangea. Such a large landmass would have a profound effect on the oceanic and atmospheric circulation on the Earth, the air and water currents taking much different paths than today. The Sun's rays would also affect the outcome: Solar radiation and surface reflection would have differed drastically, contributing to much different ocean and atmospheric circulation patterns. It was also one of the greatest periods of mountain building on Earth (due to several continental collisions) and shorter periods of more gradual uplifts, two conditions that changed the land to which most species had adapted.

All together, these factors would have a drastic effect on the climate. As proof, scientists found evidence that the late Paleozoic was a time of multiple glaciations, when the Earth's surface was periodically covered with ice. Sea level changes fluctuated in time as the ice expanded and contracted. Some scientists believe that such combinations, especially if they happened in a, geologically speaking, short period of time, could force the extinction of many species because the organisms could not adapt to the changes.

There were other changes: During the Permian, there was an unusual proliferation of evaporite deposits that increased the amount of salts in the oceans and may have caused various species to die out. Besides the glaciation, there were parts of Pangea that were hot—the equator stretched across the middle of what is now known as North America—and the climate may have become too hot for many species to survive. Still another theory states that the constant flux of sea level created shallow water environments, then dry environments; if one of the dry environments lasted longer (in tens of millions of years as opposed to millions of years), it could cause stress on the existing organisms.

Perhaps the extinctions had to do with disease. Proponents of the "elimination by disease" scenario point to the rapid rate at which our own species is susceptible to the spread of disease (often pointed out is the spread of AIDS, or epidemics, although the worldwide spread of the disease is more a case of our mobility around the planet). If such an epidemic were to spread among a certain species and kill them off, the predators on that species would be affected, and this would continue all the way down the food chain.

Other times of mass extinction were also marked by up-heavals in the Earth's natural processes. Most are associated with mountain building, others with extensive volcanic activity. Again, the Permian was a good example: In 1995, scientists Mark Richards of the University of California, Berkeley, Paul Renne of the Berkeley Geochronology Center, and others proposed that volcanic eruptions lasted a million years and flooded Siberia with lava a mile deep during the Permian. The event was potent enough to kill 96 percent of all species (about 90 percent of all marine life, 70 percent of all land vertebrates, and most terrestrial plant life). The lava not only covered an extensive chunk of land, but the cracks in the surface that released the magma from deep in the Earth also filled the sky with chemicals, dust, and gases. The fissures apparently belched out up to a cubic mile of lava each year for about 1 million years. (Of course, one may add into the mix the cause of the volcanic eruptions: Were they due to natural forces on Earth or precipitated by impacts from space?)

Were all the extinctions caused by the Earth's natural up-heavals, or were there outside interlopers that either started some of the problems or exacerbated the existing ones? The debates continue.

So far, our own species has never been through a period of extinction, although if a good-sized chunk of asteroid were to strike the planet, we might find ourselves in such a predicament. As a species, we have only been around for perhaps several million years. The reasons our entire species was never eliminated as it evolved included humans' ability to adapt, their early nomadic nature, and our proliferation as a species. Pockets of hominoids could stay away from other groups, staying the spread of disease, and thus surviving an outbreak. But if a disease spread through other types of species, they may not have been as lucky, especially if the group is stressed not only by the disease but also by a climate change.

It is almost easy to see humans recover from spreading disease, volcanic outbreaks, and general Earth fluctuations. But if a large enough asteroid were to strike the Earth, would humans have to be added to the list of extinct species?

CRETACEOUS–TERTIARY DEBATES

Not everyone agrees that such distinctive extinctions were caused by fluctuations of our own Earth's temper tantrums. And what else could cause such a wide-brush stroke elimination of species? Asteroids and comets, of course.

In 1980, Luis Alvarez, Walter Alvarez, F. Asaro, and H. Michel, scientists at the University of California, Berkeley, proposed another theory of extinction: In their now classic paper, "Extraterrestrial Cause for the Cretaceous–Tertiary Extinction," the scientists cited evidence for extinctions caused by impacting bodies.[1]

They proposed that extinction occurred because of an impact (or more logically, impacts) of a space body on the Earth's surface. This resulted in the extinction of about 50 percent of all species (including the dinosaurs) 65 million years ago at the end of the Cretaceous period, also referred to as the end of the Mesozoic era. As evidence, Alvarez and the others found the element iridium between two specific layers of rock that were laid down about 65 million years ago. The boundary between the two rock layers, known as the Cretaceous–Tertiary boundary, holds the noble metal iridium, an element that is usually associated with bodies from space.

Many scientists now agree that an impact (or many impacts) was responsible for the demise of the Cretaceous creatures. If it were truly one impact that caused the problem, the Berkeley scientists suggested that the asteroid responsible for spreading the fine layer of iridium around the world would be about 10 kilometers, slamming into the Earth at speeds close to 40 kilometers per second. (There are other estimates, too, including the same size asteroid hurtling through space at about 15 kilometers per second, 100 times the speed of the average bullet.) If there were more than one asteroid that created the chaos, they would have had the equivalent of the 10-kilometer impactor in force and destructiveness.

But no matter what the scenario, everyone seems to agree that the blast would have simulated the force of 100 billion tons of TNT, with the resulting hole nothing short of devastating, estimated at about 150 to 200 kilometers across.

The impact would kill off countless numbers of species, not only at the impact site, but also all around the world, not necessarily the result of the impact itself, but by the indirect effects of the strike. The initial result would be the immediate, extreme heat (no known experiment could come close to the actual impact of a huge asteroid striking the Earth). The impactor's shock waves in the form of a sonic blast would knock down any living thing under the impact site (for a good example, the Mount St. Helens' shock waves knocked down trees and everything else for kilometers around). After the impact, the "meteor shower" from the debris falling through the atmosphere immediately after the impact would fill the sky with a hot rain of material, igniting terrestrial forests and grasslands (the Cretaceous–Tertiary boundary at outcrops around the world show evidence of this firestorm in the form of soot). Acid rain from the fallout would have seared survivors, plants and animals alike. Dirt, ash, and poisonous gases from not only the impactor but also possibly from triggered volcanic activity would enter the atmosphere. And parts of the tenuous layer of ozone that protects organisms from the harmful ultraviolet rays of the Sun would be destroyed.

Cold would be next: The impactor would be powerful enough to blow an amazing amount of dust and debris into the upper atmosphere, to be caught by the prevailing winds and spread relatively quickly around the world. The result would be a natural sunblock, changing the climate on a global scale, lowering temperatures, and creating chaos on the planet by curtailing photosynthesis. The dead organisms would also create a breeding ground for all types of bacteria and viruses, producing diseases that would spread rapidly throughout the land.

Amazingly, the size of an impactor that could decrease photosynthesis on the planet would not have to be too large: According to Brian Toon and Kevin Zahnle at the NASA Ames Research Center, an impactor that is 3 kilometers in diameter (about 2 percent of the impactor thought to have caused the extinctions at the Cretaceous–Tertiary boundary) would be big enough to send dust into the stratosphere and curtail photosynthesis. And just as scary, such relatively small-sized impactors are known in the collection of asteroids that come close to the Earth and could eventually strike our planet if statistics are to be believed.

The reduction in photosynthesis would cause a breakdown of the food chain in the oceans and on land, and thus kill off many organisms. Only the terrestrial and marine animals that adapted to changes in climate, temperature, and vegetation patterns would survive. (One interesting aside: Not all animals and plants are recorded in the fossil record. It is often mind-boggling to think about all the species we have never seen—or even imagined—that have lived on this planet, wiped out because of catastrophic events such as asteroid and comet strikes on the Earth.)

We know that large impactors cause such clouds. Just witness the large black dust clouds that remained on parts of Jupiter for weeks after comet fragments from Shoemaker–Levy 9 struck the planet in 1994. Even a year later, the dust still had not settled out of the atmosphere. This scenario is also similar to the "nuclear winter" circumstances that have long been theorized. The explosion of nuclear weapons would dig out material and send it high into the atmosphere for long periods of time. Again, enough sunlight would be blocked to diminish photosynthesis, creating chaos on Earth.

THE SEARCH FOR TWO CRATERS

According to studies at the Cretaceous–Tertiary boundary carried out by astrogeologist Eugene Shoemaker and others, the first iridium layer was not alone. As previously noted, there are two such layers at certain sites of the Cretaceous–Tertiary bound-ary. Shoemaker notes that the first, geochronologically, is the fall-out layer from one main event (called the Chicxulub event, as explained), typically a couple of centimeters thick. The millimeters-thick second layer is on top of that, separated by a break in the sedimentary record. The two layers may indicate that two strikes occurred within a few years or decades from each other.

But could there be two such large strikes in such a short period of time? Shoemaker seems to think so: A Sun-grazing comet tens of kilometers across could have broken up during its swing around the Sun, creating a collection of rubble. Circum-stances, gravity, and orbits could have lead to two strikes or even more, some entering the oceans, others striking the continents.

After the Alvarez's iridium discovery, scientists began to look in earnest for a crater the right age and size to match the Cretaceous–Tertiary boundary demise. Luck did not seem to be with them; after all, there were only 130 craters (during the early 1980s—day's count is usually closer to 139) found on Earth, many of question-able age and most not large enough to cause such a catastrophe. To add to the problem, Earth is a dynamic planet: Such a strike would have a better probability of hitting the oceans than the land, making an oceanic impact more likely, and the Earth's crust under the sea is not only rapidly covered by sediment, but is also re-cycled by oceanic and continental plate tectonics. The Earth may have already chewed up the only evidence we have for a giant impact at the Cretaceous–Tertiary boundary.

Still, there was enough evidence to try for a solution: The iridium-rich clay layers in the Cretaceous–Tertiary boundaries around the world also held concentrations of trace elements, iso-topes, and other minerals first interpreted as originating in the oceans. Tucked away in some of the clay layers were small grains of quartz, crystals that carried parallel striations indicative of a powerful shock. Shock structures are often seen in quartz associ-ated with volcanic eruptions or even manmade explosions. But these flecks showed evidence of shock pressures higher than any known man-made or terrestrial process and pointed to a continen-tal strike.

It was not the shock structures alone that indicated a continental strike; it was the fact that the oceanic crust is made primarily of basalt, a magnesium-rich rock, whereas the continental crust is rich in quartz (silicates). The evidence suggested that the asteroid hit on or very near continental crust material, leading scientists to concentrate their efforts in one part of the world to find either a crater or other impact evidence. Candidates included the west coast of India and northern Siberia, at a multiringed 100-kilometer-wide basin called Popigai.

In the end, it appeared that the perpetrator of the catastrophe lay somewhere in North America. Attention shifted to two possible impact sites, both underground and in layers of rock thought to be about the correct age of 65 million years—the Manson structure in Iowa and the Caribbean. The Manson structure, as discussed before, is a nearly buried 35-kilometer crater that may be somewhat older than the 65-million-year-old age set for the Cretaceous–Tertiary boundary, but not by much. By about 1994, researchers determined that the structure was of impact origin, but was not the correct age to have formed at the boundary.

THE CARIBBEAN CRATER

While research was conducted on the Manson structure, others turned to the Caribbean. Several sites held some of the best evidence: Alan R. Hildebrand, then a graduate student at the University of Arizona, and others realized that coarse, rocky debris sites dotted the Caribbean basin (including some along the United States Gulf coast at several sites, such as the Braxos River in Texas). The scientists determined that the jumbled rock layers indicated a deposit from giant, kilometer-high, tsunamilike sea waves. A large impactor could throw out such a debris wave, which would easily spread out in all directions after impact. Though the Caribbean is known as an active earthquake site that sits at the site of several tectonic plate boundaries, the rocky deposit would not likely come from a seismic-generated tsunami, a large wave caused by earthquake activity in the oceans. In this case, the debris was more extensive and indicative of a much larger wave, one that could not be generated by the Earth's natural tectonic activity.

The second piece of evidence was discovered near Beloc, a small mountain village on the island of Haiti. In 1990, Hildebrand examined the Cretaceous–Tertiary boundary, discovering the usual space-indicators: the element iridium; shocked quartz pieces; and the minute pieces of glassy rocks thought to be from the impactor and/or its ejecta.

The evidence led scientists to the northwestern part of the Yucatan Peninsula in Mexico: Under the village of Puerto Chicxulub (a local Mayan name meaning "tail of the devil") on the peninsula's northern coast seems to be a crater. The finding of the alleged crater in the Caribbean was tied to a strange set of circumstances. Carlos Byars, a reporter for the *Houston Chronicle*, knew about a huge ring buried under the Yucatan Peninsula, based on information he received in 1981 from a local geophysicist named Glen Penfield. Penfield had worked as a staff scientist for the Western Geophysical Company of Houston, a company that had been hired by the Petroleos Mexicanos (Pemex) to conduct an airborne magnetic survey of the peninsula. In 1978, Penfield had been sent to the area to check on the project's progress and had found an arc pattern in the hundreds of long, strip-chart recordings.

As most geologists will tell you, the subterranean patterns in such recordings can be explained in many ways. In order to verify his potential findings, Penfield examined other gravity maps of the peninsula taken in the 1960s. There, on the other charts, was evidence of a subterranean arc, a semicircle that seemed to match the other gravity maps. Apparently, the complete circle feature covered about 180 kilometers in diameter, with Puerto Chicxulub at the hub. As an amateur astronomer, his first reaction was to believe the feature was an impact crater. His next reaction was to realize that the data was property of Pemex and could not be released to the public.

By 1981, the Yucatan project was complete, and the oil company allowed Penfield and his field supervisor, Antonio Camargo, to announce their findings. Reporter Byars was also at the Society of Exploration Geophysicists meeting where the announcement was made, and he heard the crater hypothesis proposed by Penfield and Camargo. But even though Alvarez and his colleagues had started an impact crater frenzy the year before, the planetary community still did not jump on the findings (see figure 1).

Based on preliminary core analysis, a layer of andesite was located about 1.3 kilometers below the surface of the peninsula. Such a finding indicates high heat, but in this case, the interpretation was that the area was a buried volcanic dome, not an impact crater. Penfield did not give up and tried to obtain a series of well cores from the site. Unfortunately, a warehouse fire three years before had destroyed the evidence. For three more years, Penfield tried to dig up any evidence of an impact from the area, and in 1990, Penfield and Hildebrand joined forces. After tracking down samples from nearby wells, some from Alfred Weidie at the University of New Orleans and fragments from another well tracked down by Camargo, the evidence strengthened. David A. Kring of the University of Arizona and Hildebrand found the tell-tale shocked quartz crystals in samples from both wells.[2]

But the evidence did not stop at well samples. In the meantime, satellite images of the site were gathered by researchers Kevin O. Pope, Adriana C. Ocampo, and Charles E. Duller. The images contained a segmented arc of cenotes (large sink holes) that followed the rim of the alleged crater, possibly caused by the natural slumping that takes place along the rim of a crater.

Was this the crater that changed life on Earth? Many planetary geologists and astronomers believe so. After all, Chicxulub is one of the largest circular, craterlike features on the Earth and, thus, is one of the only craters that could have wreaked such global havoc. Since the report of the Chicxulub, diameter readings, based on gravity data, show that the crater has concentric rings; one 1993 study indicated that the structure may be about 300 kilometers across, which would make it one of the largest known impact craters in the solar system. But more recent data from Alan R. Hildebrand of the Geological Survey of Canada and his colleagues in 1995 reported that new gravity measurements bring the size of the crater to 180 kilometers (sinkholes and sinkhole lakes that indicate the crater rim at the surface also indicate a smaller

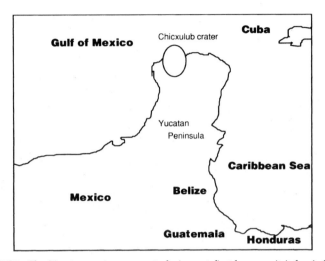

FIGURE 1. The Yucatan crater was not obvious at first because it is buried deep underground. The crater is now thought to be the giant impact event (or one of the impacts) that caused the massive extinctions at the Cretaceous–Tertiary geologic boundary 65 million years ago. (Not to scale.)

diameter). A smaller crater, maybe, but still large enough to have dealt a major blow to our planet.

The dating of the large crater has confirmed its tie to the Cretaceous–Tertiary boundary. In 1992, two independent studies used a relatively new dating technique involving the ratio of two argon isotopes extracted from the rock drilled around Chicxulub. Carl C. Swisher III at the Institute of Human Origins and his collaborators found a date of 64.98 million years (with an error margin of 50,000 years). Virgil L. Sharpton at the Lunar and Planetary Institute and his collaborators found a date of 65.2 million years (with an error margin of 400,000 years). Added to this was Sharpton's find of iridium concentrations more than 100 times greater than in terrestrial rocks, measuring about 13.5 parts per billion.[3]

IF THE TRUTH BE TOLD

Probably just as important as determining the possibility of an impact is the accuracy of our interpretation of the fossil record.

And in particular, did these extinctions, especially those of the dinosaurs, really take place at an accelerated rate? So far as many researchers are concerned, the main question is the actual time frame of extinction. Fossil evidence of dinosaurs in particular indicates that the reptiles died out gradually, not suddenly, as would be expected for such a global catastrophe as an asteroid strike.

Recent findings also indicate that dinosaurs could have existed in colder climates and that their demise was due to the natural degradation of a species. As far as we know, no species on Earth has ever existed for more than several hundreds of million years, and not every species has disappeared because of impacting asteroids or comets.

Not everyone agrees with the asteroid–comet explanation for cataclysmal extinctions. News coverage tends toward the sensational, and asteroids cutting into the Earth make for an eye-catching headline. Such coverage of impact theories has successfully drowned out others who believe that extinction was caused by other phenomena, either on their own or in addition to impacting bodies.

Still, this should not deter us from accepting the fact that asteroids do pose a threat, though, as far as we know, not an immediate one.

Chapter 1 1

BLAMING IT ON IMPACTS

Clouds, dew, and dangers come ...
WILLIAM SHAKESPEARE
JULIUS CAESAR

MOON LORE

About 10 years ago, whenever a planetary scientist proposed that the Moon formed from the Earth, the responses included strange looks or a definite debate. To say the Moon was once part of the Earth seemed almost laughable: After all, how could a chunk of Mother Earth be thrown so carelessly into space?

The idea did not originate in the past 10 years. One of the first scientists to offer a similar scenario was British astronomer George Darwin, son of Charles Darwin. He proposed the fission theory in 1879, in which a chunk of the rapidly spinning Earth was flung off to create the Moon, with the piece originating from the Pacific Ocean. Another later theory, called the capture theory, stated that the Moon was originally an independent body, simply captured as it traveled past the Earth. In a third and less dramatic theory, the Moon and Earth formed side-by-side about 4.6 billion years ago as the material from the solar nebula condensed.

The advent of supercomputer models seems to have also spawned a new model of Moon formation: Today's models seem to point to the idea that our Moon really *did* come from the Earth, albeit (to soften the blow) when our planet was in its formative stages. But why the seemingly sudden acceptance of the Moon as our offspring?

As is to be expected, as science learns more about a subject, it can add those answers to the pieces of the scientific puzzle. The origin of the Moon is one of those puzzles, and the bulk of the evidence is based on the Moon's physical and chemical constitution gathered over the past three decades. The Earth's only natural satellite (and the fifth in size of all the moons in the solar system) is, of course, one of the most studied planetary satellites in the solar system. The Moon's diameter is about one-quarter the size of the Earth's diameter, measuring about 3476 kilometers. The ratio of sizes has led many people to refer to the Earth–Moon system as a "double planet," because the mass ratio between the two bodies is much less unequal than for any other planet and their satellites. In other words, the sizes of both bodies are perfect to form a harmonious coupling.

The distances between the Earth–Moon systems are: mean distance, 384,400 kilometers from the Earth; perigee (closest) distance, 356,410 kilometers from the Earth; and apogee (farthest) distance, 406,697 kilometers from the Earth. The connection between the two bodies is evident in the dance of the Moon around the Earth. Only one side of the Moon faces our planet—meaning its rotation and revolution is the same, at 27.32 days—locked in a face-to-planet embrace.

Almost everything we know about the Moon started when the first rocks were brought back by Neil Armstrong and "Buzz" Aldrin of the monumental Apollo 11 lunar landing in July 1969. In total, chemical analysis conducted on 382 kilograms of rock and soil collected from six sites across the Earth-facing side (or near side) of the Moon has been analyzed; 0.321 kilograms of lunar material collected by three Soviet Luna craft has been checked; and 15 meteorites, collected from Antarctica and Australia and thought to come from the Moon, have been analyzed. Researchers believe the rocks came from six separate lunar sites, some possibly from the far side of the Moon.

Lunar rocks are broken down based on the two specific regions sampled on the Moon: rocks from dark maria (essentially, lava seas) and light-colored highland regions. Rocks from maria were found to be between 3.2- and 4.2-billion-year-old basalts (although it is thought that some unsampled maria regions may be as young as 2.5 billion years old). Scientists believe that molten lava in the Moon's early years rose to the surface, creating the iron and magnesium-rich rocks that cover about 30 percent of the near side and about 4 percent of the far side (the thicker crust on the far side of the Moon did not allow the magma to readily reach the surface).

The highlands regions contain mostly anorthosite, a low-density rock that is composed of high amounts of silicon, calcium, and aluminum. The low-density rock "floated" to the surface of the early magma, and the resulting crystallization of anorthosite created the highlands. Lunar isotope studies of the highland rocks confirm this; the rocks were dated at 4.3 to 4.5 billion years, and are original rock from the formation of the lunar crust.

The rocks from the Moon are also depleted in iron, nickel, iridium, and platinum when compared to the Earth's usual elements. Such lack of metallic elements have led scientists to propose the general lunar stratigraphy: A small iron core probably no larger than 350 to 500 kilometers, with a thick crust irregular in thickness.

In addition, a study of the Clementine orbiter data (the Clementine spacecraft was sent to the Moon in early 1994, but a mistake in a dress rehearsal for a planned flyby of an asteroid left the craft spinning and without steering fuel) by Paul G. Lucey and G. Jeffrey Taylor at the University of Hawaii and Erick Malaret at Applied Coherent Technology has determined the iron content of the Moon to within 1 to 2 percent by weight, using infrared absorption data taken by the craft near 1 micron in wavelength and a new method to interpret the data. The team found that mare basalts contain up to 14 percent iron, which agrees with the samples gathered from parts of the near side of the Moon. But infrared readings from the near side highlands and parts of the far side indicate less iron content.[1]

These results agree with the theory that the Moon's exterior was once a thick slab of molten rock that could be interpreted to come from the Earth's mantle. Unfortunately, the lunar gravity shows that the Moon's exterior was less of an ocean of magma than spotty ponds of magma. Lucey's team also analyzed the huge impact basin on the far side, the South Pole–Aitken Basin, which gave readings close to 10 percent iron—significantly different from the Earth's mantle readings of 20 to 30 percent iron (though this may prove that the Earth and Moon did not form at the same time or that the Moon formed as a thrown-off by-product of a fast-spinning Earth).

Finally, there are other interesting features of the surface discovered by studying the lunar rocks, including the Moon's lack of volatile elements such as mercury and bismuth. By analyzing the isotopes of lunar rock samples, scientists also determined the oxygen content of the Moon closely matches that of Earth, indicating that the Moon probably formed in the same vicinity. One more interesting finding is that the composition of the Moon and the Earth's mantle are relatively similar [for example, the percent of oxidized iron (FeO) in the Earth is 8 percent; on the Moon, it is 13 percent].

Why are composition, structural descriptions, isotopes, and other sundry characteristics of the Moon so important in determining lunar formation? By comparing the two bodies, scientists believe they can determine Earth's involvement in lunar formation. This is called comparison planetology.

And all these lunar clues indicate that a huge protoplanet (or planetesimals; some call such objects the precursors to our now-tiny-in-comparison asteroids), probably about the size of the planet Mars, struck our Earth in its formative stages. Appropriately, it is nicknamed "The Big Whack."

About 4.5 billion years ago, the solar system was filled with huge planetesimals, growing planets, and smaller-sized chunks of debris from the formation of the system. Collisions were common among the forming planets and space objects. And more than at any other time in the history of the solar system, larger planet-sized objects, still getting used to the push and pull of their new neighbors, sometimes came precariously close to one another.

Near the growing Earth, such an event probably took place about 3.8 billion years ago. One reason why the Moon is thought to come from the Earth is that our planet was still cooling down, its "softness" making it easy for a larger planet-sized object to scoop off a section to create the Moon.

As the planet and planetesimal came closer, pulled by the Earth's gravity, a collision became evident: The alleged medium-velocity collision of a Mars-sized (Mars is roughly half the size of the Earth) differentiated object striking the already differentiated Earth (meaning it had already formed a young silicate mantle and iron core). According to most computer models of the event, the impactor did not hit head on, but brushed the Earth, cutting through the crust and into the mantle. H. Jay Melosh, planetary scientist at the Lunar and Planetary Lab at the University of Arizona, Tucson, states that small impacts cannot inject material into orbit. It is a simple consequence of Keplerian orbital mechanics: To eject material into orbit, either the impact event must involve a substantial phase of hydrodynamic expansion, in which vaporized ejecta expands to a large fraction of a planetary radius before condensing into solid particles, or the impacting planetesimal must be so massive as to make major perturbations to the target planet's gravitational field.[2]

Thus, the impactor that struck the Earth had to be of sufficient size to eject the material into space and then condense, in addition to the correct velocity and angle of strike. In general, most computer models, including those by Melosh, show that after colliding, the planetesimal's core spreads out, then coalesces again, looping around the Earth only to eventually recollide and merge with the Earth's core. The mantle material of the impactor vaporizes, then eventually condenses into a orbiting proto-Moon. Such a scenario would explain why the Moon's material contains little iron and virtually no volatile elements, along with its resemblance to the Earth's mantle. It may also explain why the Earth and Moon are locked in synchronous orbit, a "double-planet" caught in each other's tidal influences.

As expected, this Earth–Moon formation scenario does not answer all questions about the Moon's origin. Most of these questions will have to await more missions to the Moon, when we may be able to determine if there are any geochemical signatures of impacting bodies on lunar rocks or to dig deeper into the lunar soil to find even more answers. And if the lunar impact theories are correct, score another point for the effects of asteroids in the solar system.

EARTH'S ENCOUNTERS WITH ASTEROIDS

The early solar system's history was dominated by collisions between objects (planet-sized and smaller) that were the major processes forming the landscapes of planetary bodies. Such collisions may have also been responsible for initial spin rates, obliquities, and satellites of the solar system. The smaller impacts would merely add or remove material from the target planet; larger impactors would form the satellites, such as described earlier with our own Moon, and encourage unique spin rates on the larger planetary bodies. Large enough impactors at the perfect collision angle could even cause strange obliquities in the planets. In fact, some scientists believe that Uranus' strange 82-degree tilt to the ecliptic was caused by just such an occurrence.

Understandably, scientists have concentrated their efforts in detecting possible results of impacts on the Earth. After all, we have direct access to the information. The fossil records reveal many drastic changes in the past on Earth, not only to organisms (mostly extinction), but also through physical changes in the Earth's crust and oceans or the Earth's atmosphere. In this case, the drastic changes may have been caused by the entry of asteroids and/or comets striking our planet.

Take for example the early Earth, which was not the most pleasant place for oxygen-breathing beings, such as humans, to be (there were still a few billion years of evolution before we arrived on the scene). The primordial atmosphere about 3.5 billion years ago was probably a dense blanket of carbon dioxide, a gas liberated by volcanic activity and by the impact of meteoric material. In other words, asteroid- and comet-type objects would tear through the atmosphere, smash into the planet, and rip open the recently formed crust of the Earth.

The exposed area would release the gases from the molten material under the crust, most of which contained a heavy portion of carbon dioxide. Along with the gas from the planet would be gas from the space objects, the vaporizing bodies also releasing quantities of carbon dioxide into the air. Apparently, there are indications that carbon dioxide did fill Earth's early atmosphere. Scientists have measured the amount of carbon dioxide in modern carbonate rocks, which, if the gas locked in the rocks were released, would not only increase the amount of carbon dioxide in the atmosphere, but would also increase the atmospheric pressure by 60 times. They also point to the atmospheres of Venus and Mars, both of which are primarily carbon dioxide, measuring about 97 and 95 percent, respectively (modern Earth has 0.031 percent carbon dioxide in its atmosphere). Since both planets are mostly carbon dioxide and are close to the Earth, then it is possible that the Earth's early atmosphere was similar.

What would the early world be like if the levels of carbon dioxide were, say, 10 to 20 times higher? The average temperature of the Earth's surface would increase, perhaps to as high as 30 degrees Celsius, much higher than the modern Earth's average surface temperature of about 20 degrees Celsius.

There is another facet to major impacts and atmosphere: If the initial atmosphere were very thin, huge impactors could have stripped away parts of the atmosphere, thus adding changes to the worldwide climate. For example, according to Ann Vickery at the Lunar and Planetary Lab, such a scene may have taken place during Mars' early history. She notes that a rain of projectiles during this time may have helped to build the atmosphere of the red planet, but then, as the planet grew in mass and increased in gravitational pull, the impacting bodies struck at higher velocities. These strong hits could have torn away a tenth of 1 percent of the atmosphere at a time, the gravitational pull of the planet unable to keep the air from escaping into space. (Earth and Venus are about twice the size of Mars, which may be why they were able to eventually keep their atmospheres.)

CRACKING THE EARTH

Could the smashing of huge asteroids into the Earth also have contributed to cracking the Earth, or was the Earth's past physical geometry controlled by giant impacts? The Earth is composed of about six large and more than a dozen small crustal plates, chunks of the crust that move around the planet in different directions and at different speeds. The theory of plate tectonics states that the plates interact with each other over time, with all of the crustal plates changing position over billions of years (see figure 1).

Is there any evidence that large asteroids and comets have struck the Earth with enough force to cause the crust to move? Most of the computer models of the crustal conditions during the Precambrian (from about 3.2 billion to 600 million years ago) show certain gaps in crustal events. Australian scientist Andrew Glikson suggests that what is missing are the consequences of mega-impacts by large projectiles, which he predicts using the lunar impact history, the Earth's early impact rates, and the modern asteroid flux. He speculates that many of the massive movements of the crust during these billions of years could have been precipitated by the collision of the Earth with huge asteroids and comets.[3]

Moving further up in time, during the Permian period about 250 million years ago, the crustal plates existed en masse in a supercontinent called Pangea. By 180 million years ago, the supercontinent had broken into Gondwanaland (or Gondwana), a continent composed of today's landmasses of South America, Africa, Australia, India, and Antarctica; a second huge continent, Laurasia, included today's North America and Eurasia. Scientists have often cited that the actual impetus for Pangea's breakup is a mystery. What would cause an entire landmass to suddenly, in terms of geologic time, break up? Enter the asteroids: Perhaps the culprit(s) was the impact of an asteroid on the Earth, splitting an already weak seam between the continents wide open. It may seem quite a coincidence for an asteroid to strike just in the right place. But if the impact were large enough, its affects could ripple through the Earth, exacerbating an already fragile fissure nearby between the continental plates and breaking the earth open even wider.

MAGNETIC, GLACIAL, AND OCEAN MYSTERIES

The Earth's magnetic field may have been caused by impacting bodies. The shape of the field suggests the Earth's core holds a powerful "bar magnet." The actual mechanism behind the mag-

FIGURE 1. Could a huge asteroid have caused changes in the Earth's crust? It is possible; the cracks in the Earth, such as the San Andreas and Garlock faults along the coast of California, would be perfect for an asteroid to rip apart. (Photo courtesy of NASA)

netic field is debated, but one theory involves the Earth's core. The field is thought to be generated by eddies that are driven by heat released by radioactive elements in the electrically conductive (and iron-rich) outer core, the entire Earth acting like a huge dynamo.

By analyzing certain layers of sediment (the evidence is recorded in the magnetic minerals found within rock layers), scientists have discerned that the Earth's magnetic field has existed since the planet's formation. But it has never been stable. In fact, as we have noted, the magnetic field has reversed itself dozens of times over the billions of years of the Earth's existence. One millennium the north pole is the north pole; the next, the north pole becomes the south pole. The change is slow, taking place over thousands of years. During that time, the magnetic field is virtually absent. (The effects of such a magnetic lapse is unknown, but several researchers suggest that life-forms that depend on the field for navigational purposes, such as migrating birds, would have problems.)

The reasons for the magnetic reversals have long been a mystery. Indeed, why should there be a magnetic field at all? Some suggest changes in the convection of the Earth's mantle or outer core. Others have blamed the reversals on asteroids and comets— the larger projectiles creating havoc if pitched just right at the planet. When the dust settles, could an asteroid or comet that punches deep enough into the crust affect the mantle enough to start a magnetic reversal?

And what about the Earth's periods of glaciation, when the Earth cools down sufficiently to create massive ice sheets that spread from the poles toward the lower latitudes? Could asteroids, again, be the culprits? About 200,000 years ago, the Earth experienced the beginning of an ice age, with four major advances and retreats of the ice sheet. Only about 10,000 years ago, the great ice sheets that spread across what is now the northern United States retreated to their present positions. (Ice, in the form of ice sheets and glaciers, now covers about 15 million square kilometers of the Earth, with the majority of ice being in Antarctica, Greenland, northern Siberia, and parts of northernmost Canada above the Arctic Circle.)

In order for the ice sheets to begin the formation process, something has to cool the Earth sufficiently. Theories include the long-term variations in the Earth's orbit (proposed in the 1920s by Yugoslav scientist Milutin Milankovich); periodic increases in volcanic activity, sending dust and gas high into the atmosphere and blocking out warming sunlight; natural changes in the Sun's energy output; and fluctuations in the Earth's own radiant heat from the natural decay of radioactive elements.

And of course, there are the theories that involve impacting asteroids or comets. Much like the "nuclear winter" scenario presented in the asteroid-extinction theories, the perpetration of a glacial ice age is thought to start in the same way. A large body (or bodies) strikes the Earth, sending dust and gas high into the stratosphere and blocking sunlight. With a large enough asteroid, the results would be global—the perfect catalyst to send the planet plummeting into another ice age.

While we are blaming everything on the impacting asteroids and comets, let us not forget the oceans. Could these bodies have added water to the planet, filling, or at least helping to fill, the Earth's oceans? Picture hundreds of asteroids and comets, fresh from the formation of the solar system, striking the Earth. Also visualize the objects rich with volatiles, such as water and methane. After all, the asteroids and comets were relatively new and had not been exposed to billions of years of the Sun's volatile, diminishing ultraviolet rays. As the objects struck the Earth, the heat from the impact, and possibly the heat from the newly forming planet, could have caused the release of the volatiles into the surrounding environment. Such a liberation could have helped to fill the oceans with water and other asteroid- and comet-associated elements.

COSMIC CONTAMINATION

Asteroids and comets are also regarded as transport systems that possibly brought the organic building blocks of life to the Earth, probably occurring early in the Earth's history.

The classic theories of life's beginnings began in 1953, when chemists Stanley Miller and Harold Urey developed a mixture of molecules thought to be similar to Earth's atmosphere at the age of 4 billion years and subjected it to a catalytic event. When heated to 100 degrees Celsius, the combination of water vapor, hydrogen, methane, and ammonia was subjected to an electrical discharge for about a week. Amazingly, four major organic molecules were generated—amino acids, nucleotides, sugars, and fatty acids in their simplest form—possible precursors to the more complex molecules essential for life. Miller and Urey's experiment was later confirmed, though with modifications, by Melvin Calvin and Sydney Fox. In 1979, Allan J. Bard and Harald Reiche produced amino acids by exposing the same solution to the Sun's rays, but they also added particles of platinum and titanium oxide.

Scientists explained that the pool of primordial soup contained the molecules filled with the precursors to life. From there, the slime that oozed in the pond eventually spawned life.

More recently, with thoughts of asteroid and comet impacts as a major process in the solar system, some scientists question the idea that life began only in shallow pools and brackish ponds, springing from the primordial soup. They argue that the Earth's atmosphere did not contain as much methane and ammonia as first theorized. And the probable carbon dioxide atmosphere was not conducive to producing organic molecules.

Many researchers now claim that life may have begun more than once, that the precursors to life may have been brought to Earth from space, or that at least life developed faster than most scientists thought possible, all thanks to impacting bodies on the Earth.

The suggestion that life on Earth came from space (or at least that the impact of asteroids or comets helped precipitate life) is not new. One of the first proponents of this idea was Sales-Guyon de Montlivault who, in 1821, suggested that seeds from the Moon were the source of early Earth life. In the late 1800s, William Thomson (Baron Kelvin) suggested that life arrived here from outer space, perhaps on meteorites. About 1905, Svante Arrhenius, a Swedish chemist, proposed the theory of panspermia, that spores of bacteria could travel great distances on the surface of objects in the cold of space. The organisms would eventually enter the Earth's (or any other planet's) atmosphere, settling into their new life on a new planet.

Most scientists agree that Arrhenius' panspermia theory is improbable, as the spores would die from exposure to deadly cosmic radiation. But some scientists do think it is possible that comets and asteroids may have helped precipitate life on Earth. The idea that comets and asteroids carry organics strengthened when, in 1986, numerous spacecraft flying by Comet Halley showed that the object contained far more organic material than anticipated. In addition, spectroscopic analysis of asteroids has shown the presence of carbon compounds; carbonaceous meteorites that fall to Earth also have carbon; Titan, Saturn's largest moon and the only one with an atmosphere, has carbon compounds; and even interstellar space has evidence of carbon—scientists have identified about 65 molecules as organic and containing carbon (for example, hydrogen cyanide and formaldehyde). It is now thought that the average organic-containing asteroid is about 3 percent organic material, whereas comets, based on the encounter with Comet Halley, are about 20 percent organics.

But just how would the organics make it to the surface? In 1989, researchers from Cornell and Yale universities proposed an idea that would allow organics, from comets in particular, to survive the grueling, fiery entry into the Earth's atmosphere. It is true that most of the organics, especially on the surface of the comet, would vaporize as they traveled through the atmosphere or would be destroyed as the object struck solid rock, no matter what velocity. But if the comet struck the oceans, and the Earth had an atmosphere 10 to 20 times more dense than it is today—a feasible scenario 3 to 4 billion years ago—it would exert an aero-braking effect on the incoming body. If the comet's velocity slowed to about 10 kilometers per second, a substantial fraction of its organic material would survive, primarily by being blown away from the impact point before there was extreme impact heating. Such work shows that it was likely that the elements essential to life (including water, certain gases, and organic molecules that include carbon, hydrogen, etc.) could have been brought to early Earth by comets and/or asteroids, which helped life to evolve.

Carl Sagan and Christopher Chyba, in 1992, proposed that there were two problems in the synthesis of complex organic chemicals from impactors: First, to identify the sources of the raw materials used, and second, to identify the source(s) of energy required in the synthesis. They claim that there was a steady drizzle of small, organic-rich particles drifting down to Earth from cometary debris (particles which would carry space-synthesized amino acids) and that seemed more likely to be the chemical precursors of life. The atmosphere is also a potential source of these chemicals, as long as energy sources are also available, such as lightning, ultraviolet radiation, and the shock energy from meteorite, asteroid, and comet impacts. Sagan and Chyba claim that such energy could have feasibly synthesized thousands of tons of complex organic compounds each year.[4]

In addition to possibly helping life progress on Earth, asteroids and comets in the early solar system may have had a different role—to negate life. Many scientists believe that large asteroids and comets in the early years of Earth's history had a major effect on life's evolution. For example, two NASA scientists at Ames Research Center in California, Verne Oberbeck and Guy Fogelman, determined that the maximum time available for the chemical evolution of Earth's life was 165 million years. They anticipate that life may have actually formed in as little as a million years after the formation of the planet—a much lower number than the billion-year time period usually cited in past life studies—but also died away almost as fast as a result of massive impacts. By using size, age, and distribution data from the Moon, the scientists extrapolated the data to the Earth, calculating the maximum time between medium-sized asteroidal impacts. Such bodies probably struck the Earth traveling at speeds close to 64,360 kilometers per hour, a force that could burn off the top layer of entire continents and boil away about 305 meters of the oceans. The life in those regions would be destroyed, perhaps leaving only life in the deepest oceans to survive to start the sequence again.

Other scientists, including Kevin Zahnle and others at Ames

Research Center, believe that life could have been killed off completely several times over by "superimpacts"—collisions between the Earth and miniplanets (the size of California) left over from the birth of the solar system. These massive impacts could have vaporized the Earth's early oceans, melted the upper crust of the planet, and erased any life that had started during the planet's first billion years.[5] Still others scientists believe that there were ecodisasters every 100,000 years, especially at the beginning of the solar system during the heavier bombardment stages.

In fact, scientists Kevin Maher and David Stevenson of the California Institute of Technology once labeled it as the "impact frustration of the origin of life," in which impact after impact eradicated the fragile beginnings of life on the surface.

Beside the asteroid–comet impact theories, there are many others, including one that says life originated at the undersea vents, a place where submarine hydrothermal vent systems carry enough warmth and elements to satisfy the earliest organisms. The living ancestors of all life forms, which are found at today's vents, display a great variety of strategies in order to survive at high temperatures. Could life have started at the deep vents in the oceans, protected from the impacts that were no doubt going on above? Perhaps the greatest question is how did life evolve from organic chemicals to the simplest life forms? Or has life always existed?

OTHER ACCUSATIONS

Asteroids are thought to be the cause of occurrences farther out in the solar system and beyond. The early solar system was filled with large impactors, rocky leftovers, as planets began to form from the solar nebula. At this time, impacts became very important in terms of rotation and axial tilt of the planets. In fact, in general, such impacts could be said to be responsible for the majority of the rotational aspects of all the planets. As the planets grew larger by gathering the leftovers, they "accumulated" the impact velocities, translating it into rotation.

Some scientists also believe that larger impactors were responsible for the ultimate rotation of a planet. At one time during its formation, Venus may have spun in the "correct" rotation, similar to the other planets. A major collision with a huge asteroid could have changed all that, creating the opposite-from-normal rotation (or maybe all the other planets were struck by larger objects, changing their rotation; but based on the overall spin of the solar system, Venus is the actual oddball). The axis of a planet could have changed in much the same way, the large impactor turning the planet away from an up-and-down axis line. Uranus could be the ultimate result of this, its axis pointing about 90 degrees from normal, making the planet rotate like a billiard ball as it spins around the Sun.

In 1991, astronomer David Andrews of the Armagh Observatory, Northern Ireland, pointed out that infalling comets or asteroids might cause stellar flares detected on young flare stars—low-mass, faint stars that suddenly brighten from the impacting space objects. The most common theory is that energetic solar flares are released from the star, but Andrews believes that episodes of massive bombardment, similar to the Late Heavy Bombardment of our early solar system, also occur in other young stellar systems. He sights the archival evidence of the star Gliese 182, a nearby M dwarf star in Orion, which flares up every 15 hours on the average, postulating that the outbursts come from the violent entry of asteroids and comets into the star's atmosphere and interior, with the star's reaction creating an observable flare.

It is obvious that the asteroids (and comets) in the early, middle, and present solar system are to blame for many events in the solar system and beyond, which is one more good reason to watch the objects that come close. But first we have to notice them to watch them. And the search has merely begun.

Chapter 12

NOTICING NEAR-EARTHERS

*The meteorites of 1908 and 1947 had struck
uninhabited wilderness; but by the end of the
twenty-first century there was no region left on
Earth that could safely be used for celestial target
practice. The human race had spread from pole to
pole. And so inevitably....*
ARTHUR C. CLARKE
RENDEZVOUS WITH RAMA

NEAR-EARTH EXTREMES

Did you ever want to get rid of a planet? In your next assignment to do just that, try a mass driver, an instrument of destruction often used in science fiction novels—a perfect vehicle for you, as the "dominant species," to get rid of an unsuspecting (or disobedient) society on a target planet. The driver delivers what it promises—mass—by sending a huge projectile through interacting magnetic fields produced when electric currents flow though coils of wire. As the mass passes through the field from one coil to the next, it accelerates. A multitude of 1-kilometer (or greater) masses hurled toward the planet would reach phenomenal velocities in only seconds. The resulting injury and destruction to the planet and its people would be difficult to imagine, as the masses drive into the surface and splash debris high into the atmosphere. An interesting, and messy, way to eliminate a burdensome planetary population.

In reality, it is nature that provides the mass driver. There is no man-made instrument of destruction, only gravity and orbital mechanics that have the power to send the mass toward the Earth.

The entry of masses into the Earth's atmosphere, or any of the other planets and satellites in the inner solar system, is obvious. The best examples in the inner solar system are Mercury and the Moon. Because of the lack of a thick atmosphere and crustal activity, we can see thousands of craters caused by ancient impacts. The other inner solar system members, Mars and Venus, also have evidence of many strikes from space objects, although many of the impacts have been weathered away because of the planets' atmospheres and prior volcanic activity. And on Earth, we have found more than 100 impact features scattered over the surface caused by comet and asteroid impacts.

As we have seen, orbital gymnastics closely follow the laws of physics: Occasionally, a comet will enter the inner solar system and sometimes make its way to our planet as it is pushed and pulled by our neighboring planets and the Sun. As for asteroids, they will stray from the asteroid belt, because of either a collision with another asteroid or the gravitational pull of Jupiter. A number of these asteroids have unusual orbits around the Sun, and some may even be coaxed out of the solar system. But there are also those that make their way into the inner solar system in close proximity to the Earth.

Collectively called near-Earth objects (NEOs), bodies (comets and asteroids) in these types of orbits have crashed into the Earth in the past, and their successors are potentially a threat to the Earth in the future. It is estimated that of the potential Earth-striking projectiles, about 90 percent are near-Earth asteroids or short-period comets; the other 10 percent or intermediate or long-period comets with orbital periods greater than 20 years (these types of comets spend little time close to the Earth, and thus have less of a potential impact threat).

The NEOs, as comets and asteroids, are usually classed according to their orbits. (One class of NEOs is rarely mentioned and only speculated about: those with orbits that lie entirely within the orbit of Earth. But so far, no such objects have been found, as they may be too small and fast to easily detect as they pass the Earth.)

The comets are usually classed as short-, intermediate-, and long-period comets. As mentioned before, the short-period comets have orbital periods of less than 20 years, intermediate-period comets orbit the Sun in between 20 and 200 years, and long-period comets have orbital periods greater than 200 years (and most of these, for all intents and purposes, are usually in parabolic orbits that take them out of the solar system, never to return). According to astronomer E. Everhart's determination of cometary orbits in 1967, about 10 to 20 percent of all the active short-period comets are Earth-crossing, with the potential for collision with our planet. (Add to this astronomer Eugene Shoemaker's size–frequency distribution of the short period comets he proposed in 1982. He estimated that the population of short-period comets with Earth-crossing orbits is about 30 plus or minus 10 objects larger than 1 kilometer diameter; 125 plus or minus 30 larger than 0.5 kilometers in diameter; and 3000 plus or minus 1000 larger than 0.1 kilometer diameter.)

The asteroids whose orbital motion brings them relatively near the Earth every few years are called near-Earth asteroids. They also go under several names, including *Earth-approaching* and *Earth-grazing*, or *Earth-crossing* (see below) *asteroids*, and in part or total, constitute the near-Earth asteroids.

Currently, more than 250 of these near-Earth objects have been discovered. Scientists estimate the total number is closer to 1000. Some are made up of the primitive material from the early solar system, others are stony minerals from the center of a partially or completely differentiated planetoids shattered from collisions, and still others are mostly metal (usually nickel-iron) from the cores of once differentiated planetoids. In fact, many of the near-Earth asteroids will be what we seek when we finally go into space, as the metals and primitive materials (which often carry water) would be perfect resources.

Of all the near-Earth asteroids found so far, most range in diameter from 32 kilometers to less than 109 meters. (The two largest known are the 32-kilometer 1036 Ganymed and the 23-kilometer 433 Eros.) The reason for the absence of smaller sizes is not that they do not exist; rather, such objects are difficult to spot. Besides such small objects being dimmer than larger objects, they are also more difficult to scout out in the grand view that encompasses 180 degrees of nighttime sky. Add to the equation the near-Earthers' relatively short distances from the Earth, making flyby appearances for a matter of days, and it is easy to see why it is so difficult to scan the sky for the elusive near-Earthers.

Not all near-Earth asteroid orbits are the same, leading to several divisions:

- *Aten* asteroids cross at least part of the Earth's orbit and are named after asteroid 2062 Aten, the first Aten asteroid discovered. The asteroids have a period of less than one year and usually spend most of their days inside the Earth's orbit.

- *Amor* asteroids have orbits that often cross the orbit of Mars and approach the Earth's orbit. The Amors were named after asteroid 1221 Amor (it was discovered in 1932; another Amor asteroid was discovered in 1898 called 433 Eros, but the Amor category did not yet exist). Amors are a type of near-Earth asteroid thought to have the potential to eventually work their way toward the Earth. But they probably will not hit in our lifetimes (or many lifetimes hence) because they are close enough to the outer planetary realm to be affected by the tug of the outer planets, especially after several close approaches.

- *Apollo* asteroids have perihelion (closest) points inside the Earth's orbit and take more than one year to circle the Sun. Because their orbits cross the Earth's orbit, Apollo asteroids are thought to be the most likely to collide with the Earth. The Apollo asteroids were named after 1862 Apollo, found in 1932.

- *Arjuna*, named in honor of the hero (an Indian prince) of an epic Hindu poem, is the unofficial name for a group of asteroids that apparently measure no more than 100 meters across and orbit the Sun in a nearly circular path. This group, with the first asteroid found in 1994, is still debated: Are the Arjuna asteroids actually Aten asteroids and therefore not readily deserving of a grouping of their own?

  The potential new class of asteroids was discovered by the 0.9-meter Spacewatch Telescope on Arizona's Kitt Peak, and many astronomers have differences of opinion on the smaller bodies. Astronomer Tom Gehrels of the University of Arizona notes that an abundance of Arjuna asteroids have been detected, some with circular orbits and others with elliptical orbits. In fact, he adds that the data once classified by the military (based on reconnaissance satellites) show a continual shower of such small asteroids on the Earth. But because they are so small, they burn up with little consequence to us living on the surface below.[1]

Probably the most intriguing unanswered question about the possible Arjuna asteroids is their origin: One suggestion is that the bodies may represent material gouged out of the lunar surface when larger asteroids smacked into the Moon. Another suggestion is that the small chunks of rock are fragments of comets that passed too close to the Earth (the problem with such a scenario is that the orbits should then be more parabolic, not circular). Still another idea is that the chunks of rock are merely from asteroids that struck each other. Astronomer Brian Marsden at the Harvard-Smithsonian Astrophysical Observatory also adds that there is little evidence that these asteroids all formed by the same processes, and if so, they should not be considered a near-Earth asteroid group. Future research will focus on determining the spectra of these swift-moving objects.

Within the list of near-Earth asteroids are the Earth-crossing asteroids (ECAs), those that have the most potential to collide with our planet. An ECA changes its orbit as a result of the constant long-range gravitational perturbations from the Earth and other planets over tens to hundreds of thousands of years. The result is a trajectory that is capable of intersecting the capture cross-section of the Earth's orbit; in other words, the possibility of the objects being captured into an orbit that crosses the Earth's. Earth-crossing asteroids are usually identified after many sightings of the asteroid, which leads to the determination of an accurate orbit for the object; thus, newly discovered asteroids will not be declared ECAs until their orbits are rigorously calculated.[2]

The largest ECAs are 1627 Ivar and 1580 Betulia, each with a diameter of about 8 kilometers (close to the size of the asteroid that supposedly wiped out so many species at the Cretaceous–Tertiary boundary). Smaller ECAs include 1991 BA, which passed within 0.0011 astronomical units from the Earth (about one half the distance to the Moon), and 1991 TU, which flew past the Earth at about 0.0049 astronomical units—both objects about 10 meters in diameter.

Astronomers believe that all the ECAs brighter than absolute magnitude 13.5 have been discovered with Earth-based telescopes. This means that, in terms of size, all the dark ECAs with low reflectivity (C-class asteroids are examples) and sizes larger than 14 kilometers have been detected; for brighter, and thus easier to spot objects, the limiting diameter is about 7 kilometers (S-class asteroids are examples). As for even dimmer asteroids (those brighter than magnitude 15, and about 3 to 6 kilometers in diameter), only 35 percent are estimated to have been found; 15 percent of the 16th magnitude objects (2 to 4 kilometers diameter) have been discovered; and only 7 percent of those about absolute magnitude 17.7 (diameters between 1 and 2 kilometers). In other words, when it comes to discovering the total population of ECAs, we have a long way to go.

How did the near-Earth asteroids evolve into these too-close-for-comfort orbits? The objects appear to have come from diverse places in the solar system, brought to their present destination by the gravitational pushing and pulling of the usual culprit (Jupiter), and to some extent, the terrestrial planets when the asteroid enters the inner solar system. (The composition of ECAs are similar to most near-Earth asteroids: The spectra seem to match closely those of the main-belt asteroids, and thus the meteorites found on Earth, which means many of the bodies were probably thrown from the belt by collision or gravitational tugs. Still another suggestion is that the smallest near-Earth asteroids, only about 30 meters across, are from chunks of rock ejected from the Moon as a result of gigantic impacts—but this has yet to be proven.)

The other inner planets and satellites of the solar system also have planet- and moon-crossing asteroids (MCAs). For example, moon-crossing asteroids travel through the space inside the Moon's orbit. Besides meteoroids, MCAs are the only objects that come so close to the Moon and sometimes the Earth. One such asteroid is about 10 meters in diameter, labeled 1994 ES1, which passed 160,000 kilometers from the Earth in March of that year, less than half the distance to the Moon. The other known Moon-crossing asteroids are 1991 BA, which passed 170,000 kilometers from Earth in 1991; and 1993 KA2, which passed by the Earth in May of that year, at about 150,000 kilometers distance.

POPULATION CONTROL

What is the population of the Earth-crossing asteroids? The estimates differ, as to be expected with a population of unknowns. One calculation of the number of smaller Earth-crossing populations is based on the power-law distribution, with the number at any size perhaps being proportional to the inverse square of the asteroidal sizes. In particular, of asteroids larger than 1 kilometer in size, there may be 2000 (maybe 1000 to 4000); for asteroids larger than 500 meters in size, 10,000 (maybe 5000 to 20,000); for asteroids larger than 100 meters in size, 300,000 (maybe 150,000 to 1 million); and for asteroids larger than 10 meters in size, there may be 150 million (maybe 10 million to 1 billion). Asteroids smaller than 10 meters in size (although most scientists refuse to call such small entities "asteroids" and refer to them as "meteors") are apparently of no consequence to us on Earth. One such body enters the atmosphere each year, but ends up as a very bright fireball, disintegrating before it reaches the ground.[3] So far, only 150 or so have been identified. To comprehend the progress of ECA discoveries, in 1989, only 90 were known; by 1991, 128 ECAs were listed. In recent years, the number of discoveries have been dwindling, mainly because of the lack of funding to near-Earth asteroid search programs.

It is estimated that the breakdown of the ECAs are Apollos (about 65 percent), Amors (25 percent), and Atens (about 10 percent). Only about half have permanent catalog numbers, implying that the orbits of the rest of the ECAs are still not well established.

In addition, some are considered lost because their locations were not well enough known even to predict their position in the sky; if their position cannot be predicted, searching for the object is an almost fruitless exercise.

CLOSEST ASTEROIDS

Many of us were youngsters when asteroid 1566 Icarus flew past the Earth, grabbing headlines in 1968. I was not as concerned as the adult world, and thought it was just an interesting oddity in the rush of my young life. After all, I believed the astronomers when they said it would pass by without affecting the Earth. But still, there were interesting throwbacks to the old days—some people warning that this was a signal of possible calamity, sooth-sayers saying it spelled out the Earth's doom, and even some idle gossip in the grocery lines mentioned that "the scientists were probably keeping something from us."

The great calamity was not to happen—Icarus sailed right by the Earth. But it was one of the first times the public sat up and recognized the fact, though not for long, that there were objects in space that we knew little about (see figure 1).

Since Icarus' rushed passage, there have been many more asteroids swinging past the Earth. Most pass by with no effect, like the 200- to 500-meter-wide asteroid 1989 FC, which came within 691,870 kilometers of Earth on March 22, 1989 (to show you the timing and luck involved, Earth had been in the exact spot only 6 hours before). If it had struck the Earth, it would have created a crater about 7 kilometers in diameter. Its blast would have been the energy equivalent of more than a million tons of TNT, pul-verizing a town underneath the impact. The surrounding region would have been torched, and the debris from the impact ejecta blanket thrown high into the atmosphere. The result would have been globally catastrophic.

And on the other side of the discussion, it may be these closer asteroids that we one day lasso and haul close to the Earth to use for its resources. Not that it would be easy, because of their sudden

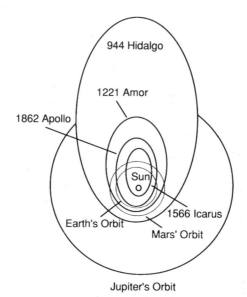

FIGURE 1. Orbits of asteroids are all over the solar system. Some travel far out into the solar system, their orbits eccentric and not a problem. But the near-Earth asteroids like Apollo and Icarus come close to the Earth. Those that cross the Earth's orbit are the scariest; they have the most potential to invade our space, possibly colliding with the Earth in the future. (Not to scale.)

appearances in the nighttime sky. But with more surveys to find the near-Earthers, such resources may, one day, be literally within our grasp.

Some of the closest approaches have occurred within the past several years, mainly because of the numerous hunts for near-Earth asteroids and better telescope technology, helping scientists find more and more of the objects:

- On January 17, 1991, an asteroid, 1991 BA, about 9 meters across, passed within 170,000 kilometers of Earth, less than half the distance to the Moon.
- Asteroid 4179 Toutatis flew by the Earth on December 9, 1992, at a distance of 1.5 million miles (2.5 million kilometers); radar images of the asteroid by astronomer Steven

Ostro of the Jet Propulsion Laboratory showed that Toutatis consists of two irregularly shaped, cratered chunks of rock, probably a contact binary asteroid. Toutatis will also make close Earth approaches in 1996 and 2000; in 2004, it will come as close as about four Earth-Moon distances (about 1.6 million kilometers).

- On May 20, 1993, asteroid 1993 KA2 whizzed past the Earth and Moon, but it was not found by Tom Gehrels with the 0.9- meter Spacewatch telescope until May 21. The asteroid, estimated to be between 5 and 11 meters across, came to within less than half the diameter of the Moon, at 140,000 kilometers distance. It traveled in nearly the same plane as the Moon's orbit.
- The asteroid 1994 ES1, at its closest approach, passed the Earth within 154,464 kilometers.
- In 1994, the asteroid 1994 XM1 surpassed 1994 ES1; James Scotti at the Spacewatch Telescope on Kitt Peak, Arizona, discovered it in December, 1994—a 13.5-meter asteroid that would pass about 101,367 kilometers from the Earth on its closest approach.

COMING CLOSE

"August 14, 2126. The end of the world?" This was the way Brian Marsden, astronomer at the Harvard-Smithsonian Center for Astrophysics (and the director of the International Astronomical Union's Central Bureau for Astronomical Telegrams) in Cambridge, Massachusetts, started an article in 1993, bringing up the possibility that Comet Swift–Tuttle would one day collide with the Earth. The prediction has since been retracted, but the story is representative of the problems encountered when scientists face the daunting problem of determining long-term movements of a small body far out in the solar system.

Comet Swift–Tuttle is an NEO and is the largest object, about 10 kilometers in diameter, whose orbit intersects the Earth's orbit. In 1973, Marsden had suggested that a comet seen in 1862 might be the same as reported by Ignatius Kegler, a Jesuit missionary, who observed the object in Beijing, China, in 1737. Marsden predicted the coincidence based on the idea that the jets from the comet, once activated on its trip around the Sun, could cause the comet slightly to change its orbit. If so, Marsden predicted the comet to arrive by the end of 1992, giving the comet a period of 130 years.

The lost comet was found in the constellation of the Big Dipper on September 26, 1992, by Japanese amateur astronomer Tsuruhiko Kiuchi. Marsden was correct, although the date of the comet's closest approach to the Sun was off by 17 days—the comet's closest approach to the Earth at 177 million kilometers. Based on the new observations, Marsden made additional calculations of the comet, noting that the next perihelion (closest approach to the Sun) would be August 14, 2126. But if the comet was only 15 days off the mark the next time it headed into the inner solar system, Marsden determined that the comet and the Earth would cross in their orbits at the very same time. Needless to say, the results would be globally catastrophic.

Marsden continued to revise the orbit of the comet, finding more and more references to the comet in records that reached back almost 2000 years (including one in 188 A.D. and perhaps one in 69 B.C.). In the end, he discovered the Comet Swift–Tuttle's orbit was somewhat stable and that the comet will safely pass about 24,135,000 kilometers from the Earth as it makes its next entrance into the inner solar system. (Marsden's orbital calculations also puts Comet Swift–Tuttle within several million miles from Earth in the year 3044. By then, we can only hope that humans will be visiting other worlds in the event of such an impact, or will have developed an efficient method to eliminate such problems!)[4]

The Swift–Tuttle story is a good example of why we need to keep track of the asteroids and comets in our solar system. What would happen if the comet were really to pose a threat to the Earth? Would we know enough about its orbit to determine if it would definitely strike us and have enough time to figure out how to cope with the problem?

More than anything else, we might want to invoke the Boy Scout motto: Be prepared.

CHAOS AND ORBITS

The fate of many Earth-crossing asteroids is not always the best for the tiny rocks. Some slip into the upper atmosphere of the Earth, turning into a bolide, a bright dart of sunlight flying through the nighttime or daytime skies, often fracturing and cracking in a fiery descent to the Earth. Most never make the ground; others make it relatively far, vaporizing just before or upon contact. One example is the 10-meter object that weighed several thousand tons and sparked and spitted its way through the atmosphere on August 10, 1972. The image was caught skimming over the Grand Teton National Park in a now-famous photo of the fireball over the mountains. The object never collided with the Earth. It burned up after lighting up the sky for 101 seconds as it traveled about 1500 kilometers at about 15 kilometers per second.

Working out the orbits of the asteroids (either main-belt or near-Earth) is not too difficult. The observer needs to fit its observed position into an elliptical path, with the Sun at the one focus of the ellipse. In principle, as we have seen with Gauss' mathematical calculations, it takes only three precisely observed positions at different times in the object's orbit to determine the ellipse and thus the orbit. In reality, it is best to have dozens of observations to verify an orbit, which helps to fine-tune the orbit and adjusts for the tug of the other bodies in the solar system.

We can even extrapolate some of the orbital data into the far future. In one case, the results seem to empty the Earth orbit corridor of the vagabonds, a tact that nature seems to favor, if the team of celestial mechanicians are correct. Several researchers, lead by Paolo Farinella of the University of Pisa in Italy, determined the orbits of Earth-crossing asteroids using numerical integrations to predict the fates of 47 asteroids. The chosen asteroids, which either cross the Earth's orbit or approach it, were sent through 2.5 million years of integration. Fifteen of the objects will strike the Sun and four will be ejected from the solar system. For example, one of the ejected asteroids will be 4179 Toutatis, to be thrown from our inner solar system in some 640,000 years. Even Comet Encke, a short-period comet thought to be the parent body of the Taurid meteor shower, will fall into the Sun in about 90,000 years.

Not all astronomers agree with such a set solution to the question of asteroid orbits. The researchers of the University of Pisa study also suggest that such orbits are highly chaotic and the results are not precise—but suggest that a large fraction of the Earth-crossing asteroids will fall into the Sun or be ejected from the system.

But what about the ones that remain gravitationally glued to the Earth? What happens to the one that gets pushed and pulled just right and becomes the newest near-Earth asteroid on the block? What if there is some quirk of fate—the perturbation of an orbit because of a passing star or the push of a shock wave from a supernova—that throws more asteroids in our direction for us to dodge?

Chapter 13

IMAGINING "WHAT IF?"

> *I do not know what I seem to the world, but to myself, I appear to have been like a boy playing upon the seashore and diverting myself by now and then finding a smoother pebble or prettier shell than ordinary, while the grand ocean of truth lay before me all undiscovered.*
> SIR ISAAC NEWTON

STRESSING IMPACTS

With all due respect to Rod Serling, for your consideration, you are an unsuspecting traveler on the spaceship Earth. Suddenly, you notice a bright streak across the daytime sky to your left. As the long streak starts to seemingly head for you and the Earth's surface, it becomes discernible as an incredibly huge, misshapen chunk of rock more than a kilometer wide. As it sinks lower, pulled by the Earth's gravity, it pushes a violent shock wave ahead of it. By now, you hear only the crackling sound as layer by layer of the rock is peeled off by the friction of the atmosphere. The shock wave hits next, but it does not matter.

Sorry. You have just been crushed by the shock wave, then vaporized by the heat of the asteroid. You and every living thing around you for a thousand square kilometers or more are dead.

And your second cousin, twice removed, living across the continent? He would not get away unscathed: The shock wave from the impact would shoot through the planet, creating numerous earthquakes along feeble fault lines in the crust, with some quakes measuring off the Richter scale. Dust and ash from the impact would spread out rapidly and dim the sky as it reached the upper atmosphere. Captured by the prevailing winds, the particles would not settle back to the surface for years. The rain of sparks would trigger wildfires around the region of the impact, adding more smoky particles to the already choked air. Poisonous gases from not only the impactor, but also from volcanic activity triggered by the impact, would permeate the atmosphere. And not long afterward, in the span of a few days to weeks, the rotting remains of the forests, fields, and towns flattened by the impact would be breeding grounds for bacteria and disease as surrounding survivors tried to cope with the aftermath of the horrendous assault. Not a pleasant scene, but feasible if a large enough asteroid struck the planet.

The above scenario is only one of many possible combinations that await us if a major asteroid or comet strikes the Earth. There are many more with equally disastrous results. And as with many problems in science, to determine the most likely effects that could await us, we must (especially in the case of large space bodies) bow to the theoretical models and computer iterations. And believe me, in such cases, no one wants to prove or disprove his or her model because the ends would not justify the satisfaction of proving the models correct. What scientists can do, however, is gather the known variables and possible related phenomena that an impactor would produce and simulate an impact on Earth. That would be just fine for the majority of us, and much better than the alternatives.

Take an object as large as the one that formed Meteor Crater in Arizona and hurl it toward the Earth's surface. The heat generated from the friction of the object's entering the atmosphere would sear everything for miles around. The rock would hit the Earth with 75 percent of its original velocity, throwing black dust high into the atmosphere and filtering the sunlight for weeks or months, possibly cooling the entire planet. Acid rain would fall, and wildfires would spread as an aftermath of the collision. The crater made by the impact would have created total destruction, devastating a small town or a part of a city underneath. There would be no time to think, because the impact would kill everyone and level all the buildings in a few seconds.

Now take an object like the 1908 Tunguska asteroid, between 30 and 60 meters across (or about half a football field), and pitch it toward a typical rural portion of the United States. The blast, thought to have been equivalent to tens of megatons of energy, would kill close to 70,000 people just from the shock wave itself, with property damage totaling close to $4 billion. In an urban area, estimates of close to 300,000 people would die the same way, with property damage surpassing more than $280 billion. In other words, a city the size of London, along with its suburbs, would be close to obliterated. And remember, Tunguska was one of the smaller incidents.

If after reading thus far, it seems likely that an asteroid strike is something to ignore, think again. Not everyone dismisses such an event as an impossibility. In fact, scientists have actually determined the average risk of an astronomical encounter of the most scary kind versus our usual risks on Earth.[1] (Figure 1)

IMPACT CATEGORIES

The relationship of a risk in relation to the size of an impacting body was studied in a workshop sponsored by NASA in 1992. The result was *The Spaceguard Survey: Report of the NASA International Near-Earth-Object Detection Workshop*, compiled by chairman David Morrison. According to the report, the hazard from a projectile can be divided into three broad categories that depend on the size and/or kinetic energy of the impactor:[2]

Category 1

An impacting body in this category usually breaks apart before it reaches the surface, and it is usually between 10 and 100 meters in diameter. The explosion of the object creates the kinetic energy equivalent to about 50 to 100 kilotons of TNT (based on a typical atmospheric entry velocity of 20 kilometers per second); overall, the kinetic energy of a body this size dissipates in the atmosphere (see figure 2).

Objects at the smaller end of the size range seem to intercept

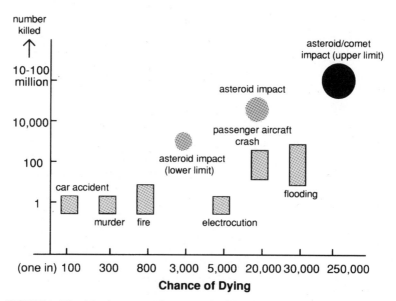

FIGURE 1. The risk of an asteroid impact is highly debated. This figure represents one theory about the chances of dying from asteroid impact and other types of accidents and the estimated number of killed per incident. This chart is based on statistics that have been based on previous impacts and current asteroid numbers, and like any statistics, it can have errors. The most important items to note on the chart are obvious: Though it is seemingly extremely rare that an upper limit asteroid will cause a global catastrophe, it would cause hundreds of millions of casualties. (After Chapman and Morrison.)

the Earth every decade. It is rare for a 10-meter impactor to survive the plunge through the atmosphere and create a crater. Only iron or stony-iron types of bodies, similar to those types noted in meteorites, would persevere, but such an occurrence is very rare. In fact, stony bodies would not survive either. The bodies would fragment as they decelerated, resulting in the chunks of rock slowing to a free-fall velocity and the kinetic energy transferred to an atmospheric shock wave. Some of the shock wave energy would be transported in a mechanical wave; the other part would appear as a burst of light and heat, usually referred to as a fireball. The explosion that results usually occurs high enough in the atmosphere to be of little or no consequence on the ground or in the

FIGURE 2. A category 1 impactor may not seem like such a problem, but the change would be considerable locally. This boulder on the Moon, found during the Apollo 17 mission, would fit in the range of category 1 impactor sizes. (Photo courtesy of NASA)

ocean. Instead, what is usually seen (and heard) is the crackling fireball and often a trembling from the shock wave as it passes by the observer.

As a body approaches the 100-meter mark, the effects worsen. Such a strike occurs on the Earth several times per millennium and equals close to 100-megatons of TNT. The actual projectile may have a better chance of reaching the ground or, because of its larger size, can actually dissipate lower in the atmosphere than objects closer to the 10-meter size. Because of this lower disruption, the energy transferred to the shock wave can also be correspondingly greater. If the threshold where the pressure in the shock wave and the radiated energy (mostly in the form of excessive heat) from the shock meet, the results can be excessively damaging. Probably the best example of such a threshold was seen in the Tunguska event in 1908, which released tremendous energy, damaging over 2000 square kilometers of the Siberian wilderness area.

What does a hit by a larger asteroid in this category mean for the Earth? The results would be local in effect, but still destructive for those around the impact site. Firestorms could rage, set off by burning embers from the asteroid and its impact. The nitric oxide and dust might be carried long distances throughout the atmosphere, although they would not affect the planet globally. Similar to a volcanic eruption, the gases and dust might create colorful sunsets for months. Physical changes in the local area if the object made it through the atmosphere and impacted the surface would be obvious—a hole in the ground and everything underneath squashed into oblivion. If it occurred over a small town or city, buildings would be flattened in a radius of 20 kilometers, and gas and oil lines would ignite from the heat of the impact. Tens of thousands could die from the crushing impact and/or shock wave or the ejecta—a natural cannonball hurling material at everyone surrounding the impact site. If the same impact occurred in the oceans, steam produced from the hot body striking the water would billow upward, causing short-term changes in the weather because of the increased amount of moisture in the atmosphere.

Category 2

Apparently, the category 2 impactors, ranging from 100 meters to 1 kilometer, strike the Earth an average of about once every 5000 years. The best possible scenario for us on Earth would be a comet of water-ice and volatiles, since scientists believe that such a body would fragment and disintegrate in the planet's thick atmosphere before it actually struck the ground. The effects of such a comet would cause localized destruction at most, mainly from the atmospheric shock bursts from the disintegrating object as it plunged through the atmosphere. This type of impacting comet would be close to the destructiveness of the Tunguska event.

But if an asteroid projectile the same size were tossed toward the Earth, we might have cause to worry even more: Metallic asteroids would probably reach and create a small impact crater on the surface, whereas a stony asteroid has to be more than about 150 meters in order to create a crater.

At the smaller end of the scale (close to 100 meters), the impactors would cause localized destruction, similar to the effects of a high-end category 1, as the energy of the impactor would be absorbed by the ground during crater formation.

For a strike by such larger impactors (closer to 1 kilometer), especially an impact on land, the splashes of ejected material would blanket an area of about 10 kilometers in diameter. Like a giant throwing a rock into mud, the ejecta would spatter in specific directions (depending on the impact angle), smothering everything it spilled on. Damage would also result from the deep crater created by the impact. This would measure about 2 kilometers in diameter and would obliterate everything that previously existed below the impact. But the damage would not be limited to ground zero: The impactor's atmospheric blast would help level buildings, forests, and most natural environments hundreds of kilometers around the area. Estimates suggest that such impacts would encompass entire states or countries, with the fatalities in the tens of millions if the strike were to occur in a populated region.

A larger, intact asteroid would also play havoc if it were to strike the oceans. The huge object would create a wall of water, sending steam and ocean water high into the sky. The result would be local changes in weather patterns (especially more precipitation), which could affect other systems on a worldwide basis. The wave produced by such a impact would be powerful enough to create a gigantic sea wave similar to a tsunami caused by seismic activity. Like dropping a pebble into a pond, the ripples would reach out in concentric patterns, possibly inundating coastlines nearby. The waves would thus cause extensive beach erosion. More devastating, however, would be its wiping out entire coastal cities and towns, killing hundreds of thousands along the densely populated coastlines.

Scientists believe that the overall effect of a high-end category 2 strike would be similar to a nuclear blast, based on the small-end megatons used in nuclear experiments over the years. But the actual effects of a close to 1-kilometer impact is pure speculation; we have no real way to determine the actual effects. Scientists can only guess, based on the localized effects of projectiles that have landed on our planet in recent years.

Category 3

One of the scariest scenarios is based on category 3 impacts. These impactors, ranging from 1 kilometer to 5 kilometers in diameter and traveling at speeds of tens of kilometers per second, were more prolific in the early days of the solar system, as seen with the many larger impact craters seen on the Moon, Mercury, Mars, Venus, and even our own planet. But that does not mean they do not exist anymore. The cratering rate of such large chunks of rock on the Earth is low—on land, they occur about once every 300,000 years (see figure 3).

How would the Earth contend with such strikes? In general, the craters produced by category 3 impactors are about 10 to 15 times the size of the projectile. For example, a 10- to 15-kilometer crater would be produced by a 1-kilometer asteroid; a 50- to 75-kilometer crater would be produced by a 5-kilometer asteroid. These are not small numbers, although they seem that way. But a 15-kilometer crater would obliterate everything comparable to the

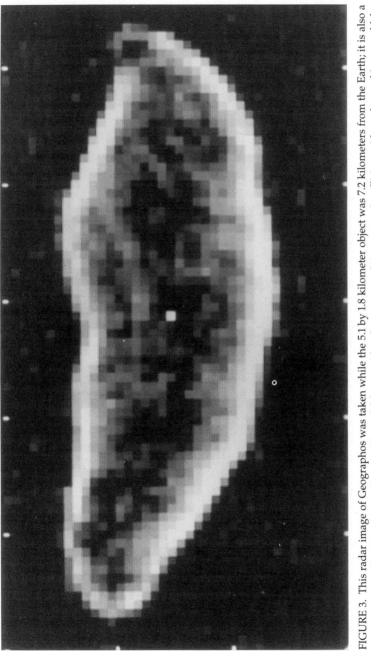

FIGURE 3. This radar image of Geographos was taken while the 5.1 by 1.8 kilometer object was 7.2 kilometers from the Earth; it is also a near-Earth asteroid. The elongated asteroid would be considered a category 3 impactor: A collision with such an object would be devastating to the Earth, changing the climate and causing major chaos around the world. (Photo courtesy of Steven Ostro)

diameter of a circle the distance between the Los Angeles International Airport and Florence, California; a large-end crater would be comparable to a circle with a diameter equal to the distance between Baltimore, Maryland, and Washington, DC. These impacts would cause local effects, but the real damage would spread throughout the world.

Although the exact size is not known, a kilometer or larger impactor would push the Earth over the threshold of a global catastrophe. If such an impact occurred on land or in the oceans the results would alter the Earth's overall balance. The crater-forming impact would disperse dust globally, enough to produce a significant, short-term change in climate worldwide, in addition to devastating blast effects in the region of impact.

So what would happen were a larger asteroid to slam into the Earth? At the smaller end of the category 3 asteroids, the devastation would be immense. And at the larger end of the asteroid category 3 spectrum, civilization itself would be threatened, if not wiped out altogether. And luckily, so far as we currently know, the chances of a collision with a larger, above 5-kilometer-wide asteroid with Earth is astronomical.

But the odds of a collision increase with the several kilometer-diameter asteroids. In short, such an impactor would first be accompanied by a massive explosion, enough to fragment and partially vaporize both the projectile and the spot below the impactor on the Earth's surface. For about a half an hour, the high-speed ejecta thrown from the impact would produce enough searing heat to scorch every living thing around the impact and would create a firestorm (from the heat and the falling fiery fragments) that would burn everything around—then spread rapidly over an entire continent. Many lakes, streams, soils, and the upper surface of some oceans would become acidified, as the nitric acid from the impactor's fireball entered the atmosphere and covered parts of the Earth's surface.

The major problem would stem from the amount of dust and debris that entered the upper atmosphere (stratosphere): Such dust would be carried around the world by prevailing winds, the dust spreading and blocking out much of the sunlight. The sunsets and sunrises would take on an amazing ruddy glow, but the dust would act like a screen during the day. The lack of sunshine would cause the temperatures to drop by tens of degrees Celsius (according to many climatologists studying global warming, even a drop of a few degrees worldwide can cause dramatic climate and physical changes, such as an increase in ice that make up the polar ice caps). This would decrease the growing seasons, or even wipe out one or more growing seasons altogether, creating massive worldwide crop failures.

It would not be over quickly. Months later, the effect would switch around—water vapor and carbon dioxide would increase, pushing a greenhouse effect over the proverbial global-warming threshold. Temperatures would rise this time, perhaps as much as 10 degrees Celsius, but such warming of the cooled down planet would probably be too late to rectify the global catastrophe. As the surface warming increased the humidity of the troposphere (the lowest layer of the atmosphere, and the one in which we reside), it would increase the greenhouse effect. Again, caught in a terrible cycle, the ocean would release carbon dioxide as it warmed, also increasing the global greenhouse warming. This thermal cycling of the planet would only strain the biosphere, decreasing the chances for all organisms (marine and terrestrial) to survive.

In the midst of the tragedy, humans would be pushed to their limits trying to survive. The colder weather after the impact would kill crops; the repercussions would include the lack of food, including loss of livestock and wildlife (who would become competitors for food in many cases) from the lack of feed. In turn, less available food would cause starvation worldwide. Hand-in-hand would be the spread of disease, not only from starvation, but also from the decay of the organisms that were killed by the impact. Fuel would be at a premium and would be in the highest demand. These resources would also be taxed to the limit as more fuel was consumed; not only physically taxed, but perhaps taxed by the leftover political structure, who would demand that we pay for the difficult to find and transport fuel resources. The unremittent daytime darkness would also increase the desperateness of the situation, the lack of warming sunshine blotting out hopes of recovery.

Months later, the warmer climate generated as greenhouse gases increased would only serve to exacerbate an already devastating situation. The warming would last for decades, creating just the opposite effect: Ice caps would melt and inundate the coastal towns and cities, and populations would move inland. The heat would put strains on the atmosphere, creating droughts or drenching rains in many areas. Humidity would increase around the world, turning much of the remaining landmasses into tropical-like jungles. The additional atmospheric water would also increase the severity of storms around the world, causing extensive flooding and damage from wind and driving rains. Humans and wildlife would again be hard-pressed to survive, fighting for the best lands and seeking places with enough food, shelter, and water. Along with the fight to win dwindling supplies would come drastic reductions in all populations of organisms.

In some cases, the Earth would recover from such changes over decades; for the even larger impactors, the effects might not be erased for centuries or could even change the geological course of the planet for the future, as was the case of the Cretaceous–Tertiary boundary (see pp. 185–187). In this case, the impact changed the course of planetary evolution, allowing the mammals to gain control of the Earth as the dominant organisms. If such a large strike were to occur today, the eradication might be mostly of mammals. And who knows which species would survive such a catastrophe to again rule the Earth.

MERELY GUESSES?

Perhaps this is not what would happen if a huge asteroid were to plow into the Earth, and luckily we have never had the experience. We are judging our predictions by the study of prior impact effects and nuclear explosions. In particular, we are basing our information on the impact that allegedly occurred 65 million years ago, which may have been responsible for eliminating more than 50 percent of the organisms on Earth, and on other extinctions that have occurred over the Earth's history.

So far, it is unknown how large the impactor would have to be in order to create mass mortality, with a significant fraction of the total number of species eradicated from the planet. Such an impact would indeed be catastrophic, but in reality, the world's population of plants and animals (including us) could suffer almost as horrendous a fate without a cataclysmic impact. After all, we are the ones who invented the nuclear bomb.

The scientists who put together the NASA categories explain that there are many variables that can throw off the drastic scenario, elucidating how they approach the risk of the larger impacting bodies:

> These uncertainties could be expressed either as a wide range of possible consequences for a particular size (or energy) of impactor or as a range of impactor sizes that might produce a certain scale of global catastrophe. We take the second approach and express the uncertainty as a range of threshold impactor sizes that would yield a global catastrophe of the following proportions:
>
> - It would destroy most of the world's food crops for a year, and/or
> - It would result in the deaths of more than a quarter of the world's population, and/or
> - It would have effects on the global climate similar to those calculated for "nuclear winter," and/or
> - It would threaten the stability and future of modern civilization.
>
> A catastrophe having one, or all, of these traits would be a horrifying thing, unprecedented in history, with potential implications for generations to come.[2]

No one can truly fathom the actual extent of a large projectile (greater than 1 kilometer in diameter) striking the Earth. The scale would be greater than the effects of all the wars on Earth packed together or even greater than five Krakatoa volcanic explosions at once. Floods, droughts, earthquakes, and other natural disasters can kill hundreds of thousands of people in a relatively short time, but a large impactor would make such hazards look meek by comparison.

The scary part is that the effects of the 5-kilometer impactor we have been discussing are small when compared to the alleged strike by the asteroid 65 million years ago at the Cretaceous–

Tertiary boundary—only just over 1 percent of the energy. Many of the near-Earth asteroids we have discovered thus far fall into the over 1 kilometer range, including the Earth-crossing asteroids 1627 Ivar and 1580 Betulia, each with a diameter of about 8 kilometers.

So what is the threshold for a huge global catastrophe? Many scientists agree with Brian Toon of NASA/Ames Research Center, who attended the July 1991 Near-Earth Asteroid Conference in San Juan Capistrano, California: The threshold diameter would have to be about 2 kilometers to create a worldwide problem of epic proportions. He also believes that the greatest harm would be created by the submicrometer dust that would reach into the stratosphere, as the extremely fine dust stays in suspension much longer than larger particles. In order to affect climate globally, the quantity of submicrometer dust is estimated at about 10,000 Tera-grams (Tg), in which $1 \text{ Tg} = 10^{12}$ grams of material; if the impactor is striking the Earth's surface at about 30 kilometers per second, the impacting body to produce such quantities of dust would only be between 1 and 1.5 kilometers.

AT ODD ODDS

If one believes that the risks associated with impactors are difficult to determine, just listen to scientists trying to determine the odds of being struck by an asteroid.

Because the Earth is covered mostly by water, there is a 70 percent chance that the collision would occur in the ocean. As mentioned, a larger asteroid or comet would cause temporary, localized climate changes by splashing a circular wall of water into the air and generating steam. If the impact were close to land, sea waves, resembling the seismic sea waves, or tsunamis would inundate coastal cities, drowning millions of people in their wake.

But what are the real chances of a large near-Earth asteroid striking the Earth? Researchers differ in their estimations. One of the best reasons for the dearth of the "we-are-next" attitudes is obvious: The Earth is a small target in the vastness of space. For such an impact to occur, it would take a good deal of complex physical interactions to get the impactor to Earth.

There are many ideas and interpretations (based on crater or asteroid size). These theories include the following:

- David A. Rothery, in his book *Satellites of the Outer Planets*, states that, "for every 10 million kilometers square (an area the size of Europe or the USA), a crater larger than 1 kilometer in diameter is formed, on average, once every quarter of a million years, and one larger than 10 kilometers in diameter every ten to twenty million years."[3]
- According to David Morrison at NASA's Ames Research Center and Clark R. Chapman of the Planetary Science Institute, in a 1990 article, a crater in the 50- to 100-kilometer range occurs on the Earth every 10 million years (with larger impacts being correspondingly rarer). Morrison also writes in a 1995 article that the larger impacts, ones caused by asteroids as small as 100 meters to ones the size that allegedly struck during the Cretaceous–Tertiary time (about 6 kilometers or larger), pose the greatest risk to humanity. He explains that even though such hits are rare, they offer the most extreme global consequences, those that would be worse for humanity than the accumulative effects of the smaller and more frequent strikes. The scary part is that he estimates there are probably 1000 near-Earth asteroids out there that have the potential to cause such global effects, and only about 10 percent have been found.[4]
- Scientists who put together *The Spaceguard Survey: The Report of the NASA International Near-Earth-Object Detection Workshop* also mention in the report that impacts of objects with diameter of 5 kilometers or greater occur about once every 10 to 30 million years.[5]
- According to calculations by Christopher Chyba at Cornell University and Kevin Zahnle at the NASA Ames Research Center, the atmosphere's air resistance puts a great deal of stress on larger objects. If the object fragments, it covers more area along the line of travel and creates extra air resistance. The result is a breakup of the asteroid, causing it to decelerate explosively. Above 15 kilometers in altitude,

there is no real harm to the Earth's surface and its people. But if object is made of tougher stuff, such as a stony asteroid over 50 meters across, the result may be an impact on the Earth with the energy of a nuclear bomb of 10 megatons. This type of explosion is thought to occur on the Earth at least once every century.

- In another estimate by *The Spaceguard Survey* scientists, the interval between globally catastrophic impacts is 500,000 years. For Tunguska-type impacts, the odds get smaller: The average interval between impacts for the total Earth is 300 years; the average interval between impacts for a populated area of Earth is 3000 years; the average interval between impacts for world urban areas is 100,000 years; and the average interval between impacts for U.S. urban areas only is 1 million years.

  They also give another list stating their prediction of the annual risk of death from impacts: For the globally catastrophic impact, the average interval between impacts for the total Earth is 500,000 years, making the annual probability of impact 1/500,000. The probability of death for an individual hangs at one quarter, with the annual probability of an individual's death being 1 in 2 million. For a Tunguska type of impact, the rates change: The average interval between impacts for the entire Earth is 300 years; assuming the area of devastation is about 5000 square kilometers (or about 1/10,000 of the planet's surface), the annual probability of an individual's death is only 1 in 30 million. And although it seems backward, it's true; the statistics show that the annual risk is about 15 times greater from a larger, kilometers-wide impactor than a smaller, Tunguska-type impactor.

- The media have a tendency to lean in the direction of chances based on the size of the impacting object. Most stories tend to carry the estimate of an asteroid larger than one kilometer across striking once or twice every million years; smaller rocks, between about 90 meters and one kilometer in diameter, hit the Earth on the average, once every 300 years. Often cited is the 1978 explosion in the

South Pacific Ocean, once considered a nuclear test, but now thought to have been a small asteroid hitting the water, and, of course, there is the Tunguska strike explained previously. Another common idea is that the overall chance of the Earth being hit by a comet large enough to destroy crops worldwide is about 1 in 10,000, similar to odds of dying from anesthesia during surgery or of dying in a car crash during any six-month interval.

In every one of these scenarios, the main culprits are near-Earth asteroids. But there is another question, one that only recently came to light, especially with the impact of Comet Shoemaker–Levy 9 on Jupiter: What if a double asteroid (or a fragmented comet) was headed on a collision course with Earth?

Certainly, the majority of asteroids we have detected appear to be singular. But it seems somewhat eerie that the second asteroid we chose to visit (Ida) had a small moon (Dactyl). In most cases, an asteroid's physical structures are determined primarily by the results of high-speed impacts. But some scientists believe that some of the smashed asteroids that circle close to the Earth may have been reduced to "flying rubble piles," while the smaller ones are probably individual pieces and may still be reasonably intact.

What would happen if asteroidal contact binaries were to collide with the Earth? Take for example the near-Earth asteroid 4179 Toutalis, the 3-kilometer in diameter asteroid interpreted as a contact binary, with 1.3- and 2.7-kilometer attached lobes. Even if the asteroids separated in some way as they approached the Earth, with the strikes on different sides of the world, the effects would be doubly devastating. No matter what the scenario, both strikes on land or the oceans, or one hit in the ocean and the other on land, the resulting catastrophe would have a devastating effect on our world.

MASS PSYCHOLOGY

Picture the headline in your daily paper: "Miles-Wide Asteroid Heads for Earth." I know I wouldn't sleep that night. Space scientists will tell you that everything within budgetary reason is now being done to search for more near-Earth asteroids and their orbits. And many of those scientists have promised that there is something we can do to find the asteroids, and plenty of time in which to do it. (The only drawback is if something were to sneak up from behind, so to speak.)

But what would happen if such an announcement were made? How long would it take before panic struck the populace? And for that matter, would such an announcement be made at all, or dropped until after the fact? What would be the responsibilities of governments who knew of the impending doom to their people? Or what is the responsibility of astronomers in making such an announcement? Would there be enough lead time to do something about the unwanted visitor from space?

There is no clear-cut answer to these questions, and every scientist and politician who would be involved in the process has his own (highly debated) reasoning. The approach, or in this case, the lack of approach, is similar to the search for extraterrestrial life. What happens if or when extraterrestrial beings are contacted? Does the government tell the public or keep the secret stashed in the depths of the "top secret" files? The encouragement that the government would tell the public stems from U.S. Congressional interest in early warning systems and near-Earth asteroid searches, but those programs, too, are now receiving dwindling support as Congress becomes obsessed with cutting programs.

The reality is there. We could face such encounters in the future. In fact, several European mathematicians recently estimated that there is a 50 percent chance that the asteroid 433 Eros (also the asteroid to be visited by the NEAR spacecraft) will become an Earth-crossing asteroid within the next 1.14 million years. And if this is true, the 23-kilometer diameter near-Earth asteroid could eventually strike the Earth—a size that would cause a global catastrophe. The researchers, from the University of Pisa in Italy and the Observatory of the Côte d'Azur in France, calculated that the object would develop a resonance with Mars; and one of the iterations showed that the asteroid could eventually cross the Earth's orbit.[6]

Right now, the best guess seems to be that there is no asteroid or comet known to be on an immediate collision course with the Earth. And many scientists say that, in the majority of cases, if a large asteroid were heading for the Earth within 5 years' time, we would have little chance to figure out what to do, except chastise ourselves for not watching the sky a little bit closer. But so long as we had more than about 10 years notice, we might have a chance. Most of us would prefer 50 years or more, a much better margin of error, allowing us to eliminate the small body before it came too close.

The estimate is that within the next century, the chance of an object 1 kilometer or more striking us is less than 1 in 1000.[7] But that 1 in 1000 could happen at any time. And it only takes one big one to ruin your day.

Chapter 14

THE HUNT AND THE HUNTED

The most beautiful thing we can experience is the mysterious. It is the source of all true art and science. He to whom this emotion is a stranger, who can no longer pause to wonder and stand rapt in awe, is as good as dead: his eyes are closed.
ALBERT EINSTEIN

POLICING THE SKY

In a now famous paper published in 1993, Eugene M. Shoemaker calculated that every year, on the average, an asteroid or a group of comet fragments comes traveling at speeds of up to 15 to 20 kilometers per second and has an explosive contact with the Earth's atmosphere. The average size of the projectiles is 10 meters, but they pack a kinetic energy equivalent of 20,000 tons of TNT. And fortunately for us, the objects are usually annihilated in the atmosphere before they can do any harm on the ground. But how do we keep track of all these possible interlopers? And how can we stay on top of the latest information to keep our "statistics of potential impacts" accurate?

Overall, the laws of orbital mechanics allow us to determine the paths of asteroids and comets that may work their way toward the Earth. But in order to determine the orbit, the near-Earth asteroids, and especially the Earth-crossers, have to be discovered. Maybe the future holds a space telescope (or telescopes) that would be totally dedicated to seeking near-Earth asteroids, but until that time, the best possible way to find the small objects is by funded searches via groundbased telescopes.

Keeping track of incoming or outgoing smaller bodies is

based on physical laws discovered as far back as Copernicus, Kepler, and Newton that say that the Earth is not the center of the universe and all planetary bodies move in ellipses around the Sun. In order to determine an asteroid orbit, we need to fit its observed position to an elliptical path with the Sun as the focus of the ellipse. Scientists usually get away with about three such observations, but many times, the three positions are not that accurately measured, so dozens or even hundreds are needed to determine the object's true path. Not only that; such iterations also compensate for the gravitational pulls from the other planets on such small bodies.

AMATEUR NIGHTS

As far back as I can remember, there always seemed to be one person at a star-gazing party who would entice us over to his or her large telescope, dial in the correct right ascension and declination, switch on the motor drive, and let us all view an asteroid. It appeared as a dot of light that actually looked like a star, but not really: Stars look like shining pinpoints and asteroids look more like tiny disks. Although the bigger attractions, such as Saturn and its rings or Jupiter and its Galilean moons, were always the "stars" of the show, asteroids were always a curiosity, another side of the solar system that few people had ever heard about or even seen.

As amateur astronomers, the view of another asteroid was another check mark for our asteroid list. We knew asteroids were merely chunks of rock, but we also knew they could tell us volumes about our solar system history. Maybe that is why so many amateurs seek out the asteroids in the sky, keeping watch over the ones that come close or continue to occupy the main-belt, and continue to tell the public all they can about the smaller members of the solar system.

A prime example is amateurs watching the asteroids during occultations, when an asteroid is "read" as it cuts in front of the light from one of the brighter stars. It is not as easy as it sounds: Stellar positions and the asteroid's path are generally well known enough to predict the object's path, but within several hundred kilometers. The occultation path, of course, is much narrower and can only be seen in a certain part of the world (almost the same idea as a solar eclipse), giving the observer perfect bragging rights if he bags it. The observer records the time when the star disappears then reappears. This can also be done with the smallest of asteroids if conditions are right, as the star seemingly winks out of sight for a few seconds. If the event is timed correctly, the data can give us a better understanding of an asteroid's shape.

Amateur asteroid hunters also contribute to the tracking, and thus the accuracy of asteroidal orbits, which is all-important in keeping an eye on main-belt and near-Earth asteroids. For example, many amateurs were recently able to track Geographos, a near-Earth asteroid that makes periodic flybys past our planet; the amateurs also search for Vesta, the third-largest asteroid, keeping track of the body and adding to the orbital database.

One master in the amateur search for asteroids was Dr. Jay U. Gunter, who, from 1971 to 1986, sent out one of the best bimonthly newsletters for amateur asteroid watchers (even professionals subscribed) called *Tonight's Asteroids*. Gunter believed in keeping amateurs informed about the various asteroids they could spot in the nighttime sky with simple telescopes. With 700 worldwide subscribers at its peak, the eight-page newsletter was acclaimed by amateurs and professionals alike as the definitive answer to those nagging asteroidal questions, especially where asteroids would be located in the sky that month.

Gunter did not run the venture full time; he was also (in his spare time, as he jokingly called it) a pathologist at Watts Hospital in Durham, North Carolina. But his work did not go unnoticed: In 1980, he was informed that asteroid number 2136 had been named after him, called JUGTA, composed of the initials of Gunter and his newsletter.[1]

Although Gunter called asteroid watching one of the loneliest of all asteroidal ventures (it is estimated that only 2 percent of amateur skywatchers are asteroid hunters), the enthusiasm for seeking out the small bodies was high then, and it is high now. There are asteroid hunters who watch one asteroid all year long and some who seek only main-belt or near-Earth asteroids. There are other asteroid hunters who seek out new asteroids for their "life lists"—keeping records of the most finds by an amateur.

Not that everyone who has a telescope can watch some of the smallest members of our system. As most amateurs will tell you, it takes at least a medium-sized telescope to get any good observations of an asteroid. Roger Harvey is an amateur astronomer in North Carolina who so far holds the record for amateur finds: As of this writing, he has already spotted 1036 asteroids.[2] Another contender in the amateur asteroid hunts is the monthly *Minor Planet Observer*, edited by Brian Warner, in Florissant, Colorado, one of the best ways amateurs can keep up with adding asteroids to their life lists. Warner produces his charts using a program by himself and the Hubble Guide Star Catalog. The results are charts of the latest asteroids, with the limiting magnitude on the charts approximately 15; each chart caption carries the asteroid's number and name, followed by the date of opposition.[3]

The United States has not cornered the market on amateur asteroid hunters. For example, in Japan, where amateur observers have long made significant contributions to the discovery and observational data of near-Earth asteroids, amateurs have been asked for more help: The National Astronomical Observatory of Japan has proposed that a network of telescopes, manned by cities and amateurs, be provided with the latest in CCD detectors to assist in the search for asteroids and comets.

As with the rest of the astronomical world, with their CCDs and computer hookups to the extraterrestrial views, amateurs are also jumping into technically advanced optics and cyberspace. As is to be expected, the more recent newsletters use the computer to generate not only star charts but also accurate positions of the asteroids over several nights (not that the previous publications were inaccurate, just more time-consuming to generate). Online services are used to exchange information on asteroids at an astounding rate. The messages are delivered in a matter of hours, as opposed to the days it takes for mail to come through. Faxes are also used, a methodology that produces information in an instant over the phonelines.

THE PROFESSIONAL HUNTERS

After reviewing all the facts and predictions about asteroids that come close to the Earth, it is difficult to accept that, over the past few years, the support for professional astronomers searching the sky for the elusive bodies keeps dwindling. In fact, according to astronomer David Morrison at the NASA Ames Research Center, there are more people working at one fast-food restaurant than there are professionals scanning the sky for asteroids.[4] Add to this poor weather conditions that can halt a nighttime search, demands on Earth-based telescopes for other astronomical projects, and telescope operational problems, and it is easy to see why many portions of the sky remain unwatched.

Many of the search programs have been in existence for quite some time. The California Institute of Technology's Palomar Planet-Crossing Asteroid Survey (PCAS) images the sky with the 47-centimeter Schmidt camera telescope (Palomar Mountain in California). Astronomer Eleanor Helin, of the Jet Propulsion Lab, became the principle investigator for the program in 1973. Films are carefully scrutinized under a stereo microscope for any objects that have moved between exposures, the tell-tale signs of near-Earthers being a dark line or trail on the photographic negative. The program track record has netted more than 45 near-Earth objects, as well as 14 comets and dozens of main-belt asteroids. This program is unfortunately being phased out but Helin and others are currently using telescopes on Maui in Hawaii to continue in the asteroid search. (Helin also started the International Near-Earth Asteroid Search in 1984, with the participation of astronomers using Schmidt telescopes in France, Denmark, Italy, the United Kingdom, and Japan. The various observatories still watch the sky for the elusive asteroids, but they, too, are feeling the crunch of funding cuts.)

Cal Tech has operated the Palomar Asteroid and Comet Survey (PACS) project since 1983, in a similar fashion as the PCAS, but depending on Lowell Observatory in Flagstaff, Arizona, for follow-up observations, and also using Palomar's Schmidt telescope; the program discovered about 43 near-Earth objects. It included work by such astronomers as Eugene and Carolyn Shoemaker and David Levy, who discovered the now-famous comet Shoemaker–Levy 9, which collided with Jupiter in 1994. Unfortunately, this search program, too, has been phased out. (Eugene Shoemaker was a geologist for the U.S. Geological Survey from 1948 to 1993 and is now scientist emeritus with the USGS and holds a staff position at Lowell Observatory in Flagstaff, Arizona. Caroline Shoemaker works as a visiting scientist at the USGS, holds a staff position at the Lowell Observatory, and is a research professor of astronomy at Northern Arizona University. Besides organizing the Branch of Astrogeology, Eugene worked on the projects to map the Moon for the Apollo lunar landings, so has been deeply involved in craters and possibilities of impacts on the Earth and other planets for years. David Levy has the status of an "amateur astronomer," although he spends much of his time at the larger telescopes, especially with Eugene and Caroline Shoemaker. Levy is described as a "writer and lecturer by day, amateur astronomer by night," but he has accumulated 21 comet-hunting discoveries, 8 of them using the 16-inch-diameter telescope in his backyard; working with the Shoemakers since 1989, they have together found 13 comets).

Despite all the canceled programs, there are several observatories that continue the search. The University of Arizona's Spacewatch survey is one such group: Members of Spacewatch include a group of astronomers working out of Kitt Peak's Steward Observatory in Arizona. Operated by Tom Gehrels, David Rabinowitz, and James Scotti of the University of Arizona, the astronomers have been searching the sky for objects, especially the near-Earth asteroids. The researchers use the 90-centimeter Spacewatch telescope and a CCD camera, the instrument including a large chip about 5 by 5 centimeters (it covers a 38-minute field of view).

The astronomers scan the same region of the sky three times at half-hour intervals. Turning off the telescopes' clock drive (used to keep the telescope in sync with the stars), the astronomers allow the stars to trail over a 2.5-minute exposure. A computer then combines the images and automatically seeks any object that has moved—the asteroids displayed as sideways streaks. Space-watch's success rate continues to be high, discovering many of the smaller, fainter (beyond magnitude 23), meter-sized asteroids that fly by the Earth. Since 1981, the program has discovered an average of two to three near-Earth objects each month, the smallest only 6 meters across. And recently, Spacewatch installed a new telescope, built with the existing 1.8-meter mirror, to search for more fainter and distant objects.

Another search program is the Near-Earth Asteroid Tracking (NEAT) program, which is also searching for asteroids and comets that come close to the Earth. NEAT, with principle investigator Dr. Eleanor Helin, is an autonomous celestial observatory at the U.S. Air Force/Ground-Based Electro-Optical Deep Space Surveillance (GEODSS) site on Haleakala, Maui, Hawaii. Started in December, 1995, it is a cooperative effort with NASA/Jet Propulsion Lab and the U.S. Air Force, using a camera and computer system on a 1-meter GEODSS telescope to search the sky for near-Earth objects.

There are also programs around the world, including the Côte d'Azur Observatory in southern France, directed by Alain Maury, who uses a new electronic detection system to watch for asteroids. The southern hemisphere is not neglected either: One in particular is the Anglo-Australian Near-Earth Asteroid Survey (AANEAS), which uses photographic plates taken through the 1.2-meter U.K. Schmidt telescope operated at the Anglo-Australian Observatory in Siding Spring, Australia. Similar to other programs, the photographic exposures are taken for reasons other than asteroid searches. The AANEAS gathers the data from the other projects, looking for the long, tell-tale streaks indicative of a near-Earth object. The longer the streak, the closer the object is to Earth, the longer the trail on the photographic plate. According to some astronomers, there are up to 250 NEOs on unscanned plates ripe for the discovering. More recently, the observatory switched to an electronic system to search for asteroids.

We cannot forget the proposed efforts of other searches. For example, the Lowell Observatory's Near-Earth Object Survey (LONEOS) is close to coming on-line near Flagstaff, Arizona. The actual telescope to be used for the search is a four-chip CCD camera mounted to a 16-inch Schmidt camera at the observatory. According to Edward Bowell, one of the organizers of LONEOS at the observatory, the detection rates should amount to 170 near-Earth objects per month, of which about 50 per month are larger than 1 kilometer in diameter (this includes those near-Earth objects already discovered); the discovery rate over a 5-year interval is predicted to be an average of 60 near-Earth asteroids per month, of which 20 exceed 1 kilometer in diameter. The final estimate is encouraging: Model calculations indicate that, after 10 years of full operation, about 60 percent of near-Earth asteroids and Jupiter-family comets larger than 1 kilometer in diameter could be discovered by LONEOS.

Not that the LONEOS technology will be lost on space. In the future, the Spaceguard effort may use the LONEOS prototype CCD technology for its own search. If the plans continue, Spaceguard will use six 2.5-meter survey scopes placed in strategic locations on Earth, allowing the system to cover the entire sky.[5]

Another near-Earth asteroid search program in the works is at the Southwest Institute for Space Research. Director Alan Hale's (one of the codiscoverers of the comet Hale–Bopp) plans are to acquire a Schmidt telescope to be based in the Sacramento Mountains of southern New Mexico. To look for the near-Earthers, Hale will use a combination of technology for the semiautomated endeavor, such as solid-state detectors and the Internet. The difference between this program and others includes participation by school students and interested members of the general public.

Still, in spite of all the efforts of these professionals, when it comes to fast-moving near-Earth asteroids, we may not be as prepared as we should be. There are near-Earth asteroids that seem to appear "all of the sudden," catching us completely unawares. Although the surprise visitors are commonly small in size and traveling very close and fast past the Earth, they still hold the potential for trouble, including a future collision.

The reasons for these fleeting flybys are simple: It is very difficult to actually detect something that has a small size and low reflectivity. The speed of such an object whizzing past us in space, and its closeness to boot, does not help. In comparison, a passing train only tens of meters away from an observer appears to speed by faster than a train farther away.

GUARDING SPACE WITH SPACEGUARD

Science fiction writer Arthur Clarke is the author of *Rendezvous with Rama*. In Clarke's novel, a program to watch the sky was instituted when a near-Earth object leveled an area of Italy. Similar to this fictional work, the U.S. Congress in the early 1990s asked scientists to organize such a committee to make recommendations on how to search for near-Earth asteroids today, before losing a good chunk of Italy, Michigan, or anyplace else in the world.

It did not matter that in Clarke's rendition the search was carried out by radar and not today's usual optical telescopic searches. What matters is that the idea is similar, and the real search was named after Clarke's fictional search, Spaceguard.

Mutterings about an internationally coordinated space watch effort probably began in 1980, after Luis Alvarez and others published their now-famous article on the cause of the demise of organisms at the Cretaceous–Tertiary boundary 65 million years ago. A plethora of papers and discussions followed, and in July 1981, NASA organized a workshop, "Collision of Asteroids and Comets with the Earth: Physical and Human Consequences," at Snowmass, Colorado. This meeting lit the match, and all of a sudden, many people became aware of the possibility of an asteroid or comet collision.

More papers followed within the decade. The next great push seemed to come in 1990, by the American Institute of Aeronautics and Astronautics (AIAA), who responded to the close passage of asteroid 1989 FC by strongly recommending a more concerted effort to seek near-Earth asteroids and determine how to prevent such objects from striking the planet. The AIAA sent these recommendations to the U.S. House Committee on Science, Space, and Technology, creating the call for a Congressional mandate for the workshop in the NASA 1990 Authorization Bill. The House of Representatives, in its NASA Multiyear Authorization Act of 1990, included the following:

> The Committee believes that it is imperative that the detection rate of Earth-orbit-crossing asteroids must be increased substantially, and that the means to destroy or alter the orbits of asteroids when they threaten collision should be defined and agreed upon internationally.
>
> The chances of the Earth being struck by a large asteroid are extremely small, but since the consequences of such a collision are extremely large, the Committee believes it is only prudent to assess the nature of the threat and prepare to deal with it. We have the technology to detect such asteroids and to prevent their collision with the Earth.
>
> The Committee therefore directs that NASA undertake two workshop studies. The first would define a program for dramatically increasing the detection rate of Earth-orbit-crossing asteroids; this study would address the costs, schedule, technology, and equipment required for precise definition of the orbits of such bodies. The second study would define systems and technologies to alter the orbits of such asteroids or destroy them if they should pose a danger to life on Earth. The Committee recommends international participation in these studies and suggests that they be conducted within a year of the passage of this legislation.

NASA's International Near-Earth Object Detection Workshop (INEODW) was the first meeting to result from that recommendation, with 24 members (all astronomers who study asteroids and comets) from four continents. The meetings were held three times throughout 1991, with the report presented in 1992. The scientists from the INEODW decided that a network of six 2- or 3-meter telescopes would detect 90 percent of asteroids a kilometer across or larger and 35 percent of the comets, giving the Earth at least 3 months lead time before the comet would strike the Earth. This 25-year search project was christened Spaceguard Survey. Along with the estimates are the usual expense accountings: The survey, which will also find objects that may become a threat even a hundred years from now, would cost $50 million for the telescopes and $10 million to operate (per year), a mere drop in the bucket by Washington standards. The plan was touted as only 4 cents for each American; with international cooperation, the amount would be less.

The second workshop on the question of altering the asteroid orbits was held in Los Alamos, New Mexico, in January 1992. The panel consisted of scientists and engineers, most of whom had been active in the Strategic Defense Initiative (SDI) program. Their approach differed from the first workshop participants; in particular, this panel was more interested in the much more numerous and frequent Tunguska-class objects, rather than the kilometer or greater class of objects. (Some scientists propose that the reason was obvious: The actual ability to track Tunguska-type objects was better suited for the advanced sensors developed by the SDI, and a great target for the special weapons introduced by the SDI program.)

One interesting proposal at the second workshop was presented by Edward Teller, the father of the hydrogen bomb and SDI, to use nuclear weapons of unprecedented yield to push the incoming comets and asteroids away from the Earth. Teller believed that such weapons, gigaton bombs with yields of over a thousand megatons (the largest ever exploded was in 1961, by the Soviet Union, at 57 megatons), should be developed and tested in space, to ensure our abilities to deflect the objects if the need arose.

Not that the NASA program was the only group concentrating on the NEAs during this time. Many smaller groups of astronomers focused on the task, and the search for the asteroids increased. International recognition of the problem also increased, with international meetings in 1991 such as the Asteroid, Comets, and Meteorite Conference (Flagstaff, Arizona), followed by the International Conference on Near-Earth Asteroids (San Juan Capistrano, California), and the "Asteroid Hazard" meetings (St. Petersburg, Russia); the International Astronomical Union's endorsement of international searches for near-Earth objects (NEOs) and their Working Group on Near-Earth Objects; the 1993 Space Science Colloquium on the Hazards of Impacts by Comets and Asteroids (Tucson, Arizona); the 1993 Erice Seminar on Planetary

Emergencies: Collision of an Asteroid or Comet with the Earth (Erice, Italy); and the Space Protection of the Earth SPE-94 in 1994 (Chelyabinsk-70, Russia).

BEYOND SPACEGUARD

By 1993, the results of the 1991 to 1992 studies were released to the House Subcommittee on Space. Subsequent to a hearing on March 23 that year, the Committee on Science, Space and Technology approved the following additional direction to NASA in July 1994 (which also coincided with the Jupiter comet fragment impacts): "To the extent practicable, the National Aeronautics and Space Administration, in coordination with the Department of Defense and the space agencies of other countries, shall identify and catalog within 10 years the orbital characteristics of all comets and asteroids that are greater than 1 kilometer in diameter and are in an orbit around the Sun that crosses the orbit of the Earth."

The Spaceguard Survey was not forgotten, just upgraded. Technology has improved substantially to carry out an even more ambitious program: The plan now permits a complete survey of the projected 1000 to 2000 Earth-crossing objects larger than 1 kilometer (although smaller objects also will be tracked), and using both NASA and the U.S. Air Force technologies, a rapid survey of near-Earth objects is in the works. The goals of the survey are obvious: First, the orbital elements of a potentially threatening object should be determined with sufficient accuracy to predict its position on its next approach to the Earth and to obtain the possibility of an impact at a future date. Second, all bodies discovered should be classified by their reflectance and spectral type to comprehend the populations of various asteroids, in terms of size and composition. This information would be useful when planning ways of eliminating a potentially threatening object.

In general, besides the continuation of the sky searches mentioned above, in particular, the Spacewatch Telescope at Kitt Peak and the Lowell Observatory Near-Earth Object Survey, both in Arizona, the recommendations of the current survey include many technological upgrades to existing facilities. For example, several scientists have recommended to use more of the Arecibo radar/radio and Goldstone radio telescopes for near-Earth object studies, as the facilities could be (and have been) used to image nearby objects. Another recommendation is also to use the U.S. Air Force Space Command's network of 1-meter wide-field telescopes (as the Groundbased Elecro-Optical Deep Space Surveillance system), now under development. Participation also extends into the international astronomical community, with suggestions on how to improve the searches with new or upgraded facilities and to extend telescope time to fit in the near-Earth asteroid search.

WE KNEW BEFORE

Data on near-Earth asteroids continue to be collected from both amateurs and professionals alike. What may be tens of thousands of asteroids, mostly small, may exist within the inner solar system, and there is a definite desire to determine the orbit of all of the rocky bodies that pose a potential threat to the Earth.

But let's take that data and ask a simple question: Just what would happen if someone, amateur or professional, were to find that an asteroid or comet heading for Earth? Could we actually do something about it?

Certainly, a smaller projectile would be of no consequence. (Estimates indicate that objects smaller than about 100 meters in diameter could be stopped by throwing something big, such as the proposed Boeing's Lightweight Exoatmospheric Projectile developed for the Strategic Defense Initiative, a huge projectile that would impact and throw off the asteroid.) A slightly larger asteroid could feasibly cause local climate and wildlife habitat changes. But a larger chunk of rock or ice and rock could wreak havoc on land or in the oceans, creating a chaos that could lead to global climate changes, such as a decrease in temperature, and even the annihilation of cities near the strike zone.

Comets would even be more interesting to encounter. While

asteroids move at about 25 kilometers per second, comets usually approach the planet at twice that speed. The rapid approach would give us much less time to counter the collision, and if the body were to make it through the atmosphere, the strike would produce just over the energy of an asteroid of comparable size. The long-period comets seem to be the most threatening: Because most of their orbits are longer than 200 years, the movement details of most long-term comets are unknown, making it difficult to determine an actual orbit.

This is not the only time in history we have been aware of such a threat to the Earth. Project Icarus, presented in 1968 at the Massachusetts Institute of Technology, was a plan to head off an asteroid that strayed too close to the Earth. The Icarus program was fashioned after the Apollo-Saturn 5 technology of the late 1960s, the capsule and launcher that took the astronauts to the Moon. If an asteroid were detected, the emergency would increase the production of a Saturn 5 and Atlas-Athena assembly lines that existed at that time. The resulting Icarus would use the two existing Saturn 5 launch facilities, and a third would be built to support the project.

Project Icarus originally used six highly modified Apollo and five altered Mariner spacecraft. Strapped to the Apollo craft would have been one 100-megaton thermonuclear warhead with six available for the project. The craft would be launched to meet the asteroid; and as the warheads detonated, the asteroid would be nudged out of the Earth's path.

Not everyone has given up on an Icarus-type project. In 1992, Daniel James Gauthier pitched a similar idea. We no longer have the assembly lines for the Saturn 5, but he mentions some alternatives, especially in the heavy-lift vehicle production lines and the Energia launch pads at Tyuratam Cosmodrome that are owned by the Commonwealth of Independent States in Asia. He suggests that the Block DM stage and Prognoz radiation monitoring spacecraft could take the place of the Apollo and Mariner hardware.

Production of the warheads is more complicated: Not only has production of warheads been shut down in the United States and CIS, but to ask for such a large warhead even during top production periods would be next to impossible (as noted, the largest reported warhead ever tested reached 57 megatons; the largest operational weapons yielded about 25 megatons). The answer may be a multitude of 10-megaton bombs (it is reported that a 10-megaton bomb has about half the blast effect of a 100-megaton bomb), pitched together into space. The number of warheads and launches needed for the asteroid rendezvous would depend on the asteroid.

The idea is to launch the warheads toward the asteroid, allowing it to strike the object and blow off material from the asteroid surface. The asteroid would then respond in the expected fashion, based on a major premise in physics: For every action, there is an opposite and equal reaction. The material exploding away from the surface would act almost like a cosmic "kick," sending the asteroid in the other direction.

And of course, the entire scenario to stop the asteroid would depend on how much notice we had before the collision. To prepare for such an emergency would take close to two years, perhaps more, depending how far in the future such a contingency exists; after all, as more and more warheads are dismantled, we would need that much more time to begin emergency production.

Knocking out a large comet would call for a different tactic: Because most comets are loose conglomerations of ice, dust, and gas, the nuclear explosion would need only to come close to the comet, exploding nearby to create heat. The nearest surface area of the comet would melt, creating man-made jets as the frozen gases melted, the jets acting like rocket exhaust engines, turning the comet away from its path.

Of course, by the time such a scenario comes into being, our technologies may well improve, creating a better method for eliminating the threat to the Earth. Then again, we may never, in the next few generations, have to experience such a dire threat to our planet. One of the reasons for such optimism is that time is (most often) on our side. A comet heading for Earth would be detected rather rapidly; thus by the time the object reached the Earth, we may have found a solution.

AND NOW?

Bringing us up to the present, NASA's asteroid workshop conferences addressed the subject of how to handle an object headed straight for Earth. The consensus was that, based on our current technology, nuclear would be the best way to nudge or eliminate an asteroid. Boosts from a Saturn 5 rocket are out of the question (the Saturn 5s were dismantled because they were no longer needed) but the Titan 2 Centaur, or the Russian Energia or Proton, have been suggested as workhorses to bring the nuclear bombs into space.

But what are the realities of such an explosion in space? How easy would it be to make contact with the asteroid and push it? How would we know that such a shove would put the asteroid careening in the "correct" orbit? Could we somehow lasso the object and maybe harness it in such a way as to create a new source of raw materials instead? Or could we really blow up a threatening asteroid (or comet) using a nuclear device? Would superpowers on both sides of the nuclear disarmament treaties confess to having 20-megaton bombs in storage? In addition, politically speaking, the 1963 Limited Test Ban Treaty and the 1967 Outer Space Treaty both ban the use of nuclear weapons in space. Not only that, what would the reaction of the populace be if we were to ignore the treaties in order to save the Earth? Every time the Space Shuttle launches a craft that is nuclear-powered, special interest groups line the launch site to try and stop the proceedings. If we were to launch 30 nuclear warheads, even if it were to save ourselves from a global catastrophe, what would be the public's response?

The solutions are not quite clear-cut, especially whether to blow up the asteroid or just nudge it off course. The problem with smashing an object with a nuclear device is a question of not only nuclear safety, but also safety from the asteroid fragments. After the asteroid is blown apart by the nuclear explosion (or explosions), the resulting rain of fairly large-size chunks of rock could have more globally devastating consequences than the strike of the original asteroid. If the chunks were large and spread out enough, their searing heat could spark massive firestorms around the world. In fact, because of this problem, some scientists suggest that they would rather ride the storm of the colliding intact asteroid and attempt to evacuate the target zone and stockpile food in preparation for the strike! Others suggest superblasting the object, resulting in small, less destructive fragments. But such an endeavor would involve knowing the enemy backward and forward in order to split all the fragments into the correct size.

Another suggestion that would eliminate the problems of fragmentation was proposed by Johndale Solem, a mathematical physicist at Los Alamos National Laboratory: to push a threatening Earth-crossing asteroid out of its collision-oriented orbit with a neutron bomb. The more gentle push from explosions off the asteroid's port or starboard bow (called standoff bursts) would decrease the chance of fragmentation. Thomas Ahrens of the California Institute of Technology and Alan Harris of the Jet Propulsion Lab estimate that a 100-kiloton standoff burst would be sufficient to change the velocity (and thus, the orbit) of a 1-kilometer asteroid by 1 centimeter per second, enough to send the threat on its way. Others propose the use of spark guns similar to those developed for the "Star Wars" initiative at Sandia National Laboratories in the 1980s. After landing on the surface of the asteroid, the guns would gather up, then throw pieces of the surface into space, gently moving the asteroid in opposite direction from the throw. Still another more elegant way to get rid of the offensive object would be to attach huge solar sails to the asteroid. The unfurled sails would catch the solar wind, which would slowly push the asteroid away from its target, the Earth.

The problem at the moment is: "What do we do now?" Do we spend money now to develop programs, so we are in readiness for a future asteroidal threat? Not everyone agrees. According to Steven Ostro at the Jet Propulsion Lab and Carl Sagan of Cornell University, the approach to the problem must be "strategically intelligent" in reducing the risks, not only from space objects striking the Earth, but also from other potentially life-threatening risks. In fact, both scientists suggest shelving any system that would defend the Earth against asteroids, as the threat is too remote at this time to justify the costs. They also note that such a defense system may be a temptation to a Hitler- or Stalin-type political leader (somewhat à la James Bond), and if we could stop an asteroid from colliding with Earth, a desperate leader could conversely threaten to push the asteroid toward the planet.

The majority of scientists agree that it is perhaps premature to develop such an intricate system, especially since there is no evidence of an immediate hazard. Right now, the best strategy should be to use our improving telescopic technology to continue to catalog new near-Earth objects and gather more data on their orbits to determine a potential threat. And while we continue to gather more data, we should develop scenarios to help us to defend ourselves against a possible asteroidal attack from above.

In other words, more funding should be granted to the search programs, our only hope if we are to keep close track of the smaller bodies in our system. We need to just keep watching.

Chapter 15

RESOURCES FOR THE FUTURE

*After a moment, Spock withdrew his head from
his computer's hood. "Typical asteroid chemically
but it is not orbiting, Captain. It is pursuing an
independent course through this solar system."
"How can it?" Kirk said. "Unless it's powered—
a spaceship!"*

RICK VOLLAERTS
STAR TREK 8, ADAPTED BY JAMES BLISH
FOR THE WORLD IS HOLLOW AND I HAVE
TOUCHED THE SKY

WHENCE FROM HERE?

Orbiting in the recesses of solar system, farther from home than the Moon, a tiny fleck of light is all that tells a human space traveler of his destination. As the reflected light alternates between bright and dark every few hours, the traveler realizes his target, an asteroid, is spinning. As the view of the minor planet becomes larger, the traveler's first impressions are typical for first contact with an asteroid: The surface is scarred with craters, deep grooves radiate from the object's largest crater, and some areas are covered with a thin layer of gray regolith, similar to the color of the lunar highlands. The asteroid is somewhat angular in shape, and has a tiny moon nearby, undetected until the spacecraft comes closer. He arrives and soon hitches himself to the asteroid, like a climber up the side of a steep precipice on Earth. It is time to look around and see what treasures the tiny object will yield, resources for Earth and humans' expansion into space.

Despite the fears of gloom and doom that asteroids pose, they also elicit rays of hope. Not every near-Earth asteroid is a threat to the Earth, and those that are not can be a virtual gold mine of much-needed resources, for the world and for living in space.

Humans have been in space for years. We have orbited the planet numerous times, sometimes for over a year at a time (Russian cosmonaut Valery Polyakov has spent the most time in space, recording 438 days in Russia's Mir space station). We have made it to the Moon and back at least eight times (though two of the flights just swung around the Moon). And we have sent spacecraft to the edges of our solar system, some successfully, some not. Finding new frontiers is in our blood. Space is a new frontier—and a way of advancing our knowledge, technology, international cooperation, and an interest in life itself.

As a scientist and amateur astronomer, I am thoroughly committed to the prospect of humans and robots in space. I have entertained such flights of fancy for myself, but in reality, my trips "into space" only means a quick flight in the troposphere, the astronauts will be the ones in real space. But if a robot can explore better and cheaper than a human, that is where we should invest when planning to visit asteroids. The constraints on humans are obvious: Space flights may entail years in space, long hours of flight to and from the target object. The wear and tear on the body is difficult to calculate. And there are questions of safety, health, and psychological states in the voids of space.

That is why our first steps into the solar system should be unmanned visits to near-Earth asteroids, especially the close Apollo asteroids, and only eventually, manned flights to the closer minor planets. Small craft would be easy and relatively inexpensive to launch from Earth for near-Earth asteroid investigation, or from the Moon or Mars for rendezvous with main-belt asteroids. The first flights will probably be small craft with new technology, giving us an idea of how to reach the object. Additional missions to more asteroids would allow us to pinpoint the potential of mining certain types of near-Earth asteroids. Such craft could examine the varying chemical composition to judge which bodies would be best for mining.[1]

TRAVEL PLANS

Getting even a small craft to the majority of asteroids within the asteroid belt would not be easy. The overall mission velocity, or "delta-vee" as it is called by the people who work out trajectories to send ships to other worlds, is much higher than that to visit the Moon or Mars. But visiting the Moon and Mars are payload, time, and funding intensive, which is why we should shoot for the easiest targets in the solar system: the near-Earth asteroids.

Many of the near-Earth asteroids that swing by the planet are called Triple-As, or close in terms of delta-vee; about six are closer than the Moon, and more than 50 are closer than Mars. With recent concentrations on seeking near-Earth asteroids, it is certain more will be found, including those perfect for a visit.

Although manned missions are an eventuality at this moment, in the mid-1960s was a time when NASA scientists visualized a manned rendezvous with a near-Earth asteroid, although it did not receive as much press as the lunar journeys. Knowing of asteroids at that time, they planned to use a two-man Gemini capsule propelled by Rover nuclear rocket stages. Later in the decade, an expedition using unmodified Apollo-Saturn 5 hardware was given serious consideration as part of the Apollo Applications Program (AAP). But plans were dropped as the funding plunged, reducing the AAP proposition to funding for the space station Skylab in the early 1970s.

Across the ocean, the then-Soviet Union proposed a manned flight to a near-Earth asteroid, to be achieved in the 1990s. An Energia launch vehicle capable of sending a Salyut spacecraft crew module would fly a crew on a rendezvous to a near-Earther, and return the crew to Earth. The worst part would be the wait to get to the asteroid. But in reality, funding and time again dwindled, and the idea was shelved. Interestingly enough, Russian crews are currently being tested for long-duration exposure and weightlessness in the Mir space station and have a head start in sending eventual manned missions to a near-Earth asteroid.

For now, manned missions to a near-Earth asteroid are considered (very) future projects. A space system to transport a manned spacecraft to an asteroid does not exist, even if the shuttle were redesigned. There is just no means of reaching an asteroid, complete with equipment and personnel payloads, at least not at our current levels of technology and funding.

CLOSE AND GETTING NEAR

That does not mean that scientists will ignore all possibilities for connecting with near-Earth or other asteroids. NASA has a policy mandating that all outer solar system missions be devised to explore minor planets along a traveling craft's trajectory. The spacecraft Galileo, on its way to the planet Jupiter, did it, quickly flying by the main-belt asteroids Gaspra in 1991, and Ida in 1994.

The past lists of proposed craft to a near-Earth asteroid (or any asteroid, for that matter) are long, with the majority of the space ventures meeting funding demises. For example, engineers at NASA's Jet Propulsion Laboratory proposed the Asteroid Inspection with Microspacecraft (AIM) that would have sent three 36-kilogram spacecraft, launched on a Pegasus/XL vehicle, to three separate near-Earth asteroids. Similar to almost everything entering the microspacecraft market, the AIM was to use lightweight sensors developed by the military for the Strategic Defense Initiative. Not that this technology is going to waste: Recently, the government has given permission to several commercial groups to build lightweight, high-resolution satellites and will allow the 1- to 3-meter resolution satellite images of the Earth to be sold commercially. This is still not up to government satellite standards, which, reportedly and understandably, have higher resolution, enough resolution from the satellite images to spot a football on the Earth's surface. But if used for future asteroid missions, the resolution would still be perfect for imaging near-Earth asteroids.

One of the missions to be cut from NASA's budget was the Comet-Rendezvous and Asteroid Flyby (CRAF), even though the planning and design had been going on for years. The craft was to encounter the main-belt asteroid Eunomia, then fly in formation with the comet Wild 2 for several years. Unfortunately, the mission was scrubbed from the roster because of NASA budget cuts in January, 1992, the first time a fully approved planetary mission had been terminated.[2] Another loss of an asteroidal visit came in early 1994, when the Clementine craft was scheduled to fly by the asteroid 1620 Geographos after mapping our Moon for 2 months. But during a dress rehearsal with the Clementine for a flyby of the asteroid, the spacecraft opened four of its altitude-control thrusters, mistakenly leaving them on for 11 minutes until the steering fuel ran out. The spinning craft is in solar orbit today, but with little or no hope of any recovery.

Such problems have not helped the promotion of space ventures. So recently, the space agency seems to be leaning toward smaller, leaner, and less expensive designs, part of a proposed series of small, low-cost planetary missions known as NASA's Discovery program. One such project is an unmanned craft to visit a near-Earth asteroid: the Near-Earth Asteroid Rendezvous (NEAR) mission. It was built and is operated by the Johns Hopkins University's Applied Physics Laboratory.

The original mission was to launch the craft to the near-Earth asteroid 4460 Nereus, in January 1998, an easy target to reach in terms of energy. But although the funding would fit well within the guidelines of the low-cost mission, scientists worried about the size of the asteroid, a mere kilometer in diameter, as a drawback to obtaining the best scientific information. The mission to Nereus was withdrawn, and today the craft is scheduled to fly by asteroid 253 Mathilde, then match its course with the near-Earth asteroid 433 Eros, a 40- by 14-kilometer, S-type asteroid, an unusual Amor-type orbit that takes it near the Earth (many Amors are thought to be extinct comet nuclei, but Eros seems to be an ordinary asteroid from the inner asteroid belt, perturbed out of its original orbit).[3]

Not that it will be easy to get to Eros. According to Joseph Veverka, space scientist at Cornell University and principal investigator for two of NEAR's instruments (the camera and near-infrared spectrometer), the Eros rendezvous will take a great deal of energy, requiring a launch on a trajectory highly inclined to the Earth's equator, meaning that the craft will not have the full benefit of using the Earth's rotation to pitch it toward the asteroid rendezvous. A larger rocket to boost the craft was not feasible, so mission designers found that a 2-year delta VEGA (delta-V, velocity, and Earth Gravity Assist) was the answer. In other words, they would give the craft a longer time to reach the asteroid as it goes through several iterations to throw the craft to the rendezvous point.

The launch took place in February 1996. NEAR's first mission will be to fly by the asteroid 253 Mathilde, which should take place around June 1997. The large main-belt asteroid is thought to be about 61 kilometers in diameter and is a C-type asteroid, the first time for such a close-up view of another asteroid type (besides Eros, the two other visited asteroids, Gaspra and Ida, are S-type asteroids).

After the challenge of getting to Eros by February 1999, there is also the task of keeping the spacecraft in working order as it rides along with the asteroid. NEAR's first approach to the asteroid will reveal the body's physical attributes: determining the mass to about 1 percent accuracy, examining its surface features, and accounting for its shape and spin rate. But the asteroid's irregular gravity field could present certain problems, especially at altitudes around 25 kilometers or under above the surface. One false move, or not pulling out in time as the asteroid's and craft's trajectories intersect, could mean a crash landing for NEAR. For 11 months, while the spacecraft gathers its scientific information, scientists operating the control center for NEAR (Johns Hopkins Applied Physics Laboratory) and navigation center (Jet Propulsion Laboratory) will be on constant standby. When the project ends in December 1999, scientists may then send the craft to land on the asteroid, the first time such a feat has been attempted.[4]

The rewards of the distant and close-up views will be invaluable. The near-infrared spectrometers will map the minerals on the asteroid's surface; the entire surface will be imaged (including resolutions at scales of 3 to 5 meters), making it possible to determine variations in minerals and spot features that can lead to the interpretation of asteroidal processes (physical and geochemical); gamma-ray/X-ray spectrometers will analyze the elements on the surface, which may lead to proving (or disproving) the asteroid–

meteorite connection; radio science measurements will allow us to "see" into the asteroid, to determine if the body is solid or just a orbiting pile of rubble, broken by collision after collision; and of course, NEAR will search for a possible moon, a phenomenon that we have recently found is more common than anticipated, or debris around the asteroid.

Not only will NEAR reveal more information about the early solar system (and perhaps gain some insight into the earliest processes that shaped the system), but it also will be a step toward eventual rendezvous with other near-Earth asteroids. And that step may one day lead us to a manned exploration of the asteroids, not only for our personal edification, but also to use as resources in space.

HITCHING TO AN ASTEROID

After the robots have finished their initial exploration, we should launch the first long-term manned flights to asteroids, especially the near-Earthers. These flights will be able to answer many of our technical and health questions in the quest to expand into space. Visits to these small bodies will also provide us with much-needed resources, including (depending on the asteroid) organic chemicals, oxygen, water, biomass, and fuel to help sustain future human space colonies. They would also provide much-needed Earth metals, including platinum, gold, and other valuable ores, and maybe some minerals we have yet to discover.

Exploring an asteroid will not be the same as going to a planet. One major difference is the size of the asteroid; thus, the gravitational pull and escape velocities are much lower. Because of this, if you tried to stand on a smaller asteroid, even one as "small" in size as the 11- by 15-kilometer diameter asteroidlike Martian moon Deimos, a quick stroll would be almost impossible (just look how the Apollo astronauts on the Moon bounced relatively high as they walked). To compare, an asteroid 10 kilometers across would have a force of gravity about 0.1 percent that of Earth. If the Earth's escape velocity is 11.2 kilometers per second, you would only need speeds of 4 meters per second to escape the minor planet's gravitational pull (the Moon's escape velocity is 2.37 kilometers per second). In other words, a pitched baseball would have no trouble going into orbit around such an asteroid.

Because of the low escape velocity, landing the mother ship on the surface would be impractical. Landing gear would be designed for emergency use only, as the crew would visit the surface via lifelines or power packs. According to Eugene Shoemaker, to work on the surface, the astronauts would fire a little bolt with a loop, similar to a mountaineer's piton, into the surface. They would then hitch themselves to the asteroid and use ordinary rock-climbing techniques to explore the surface. A battery- or solar-powered backpack could also be used to maneuver around the body.[5]

RESOURCES FROM THE PAST FOR THE FUTURE

In space, as in life, the saying "you can't take it with you" becomes readily apparent. The massive nature of the effort to build infrastructures and colonies in space has become evident as we ponder the task of building the international space station. Even the shuttle program's launches have shown us that payload is a major concern and obstacle. How many launches must we make before we can begin building a colony on the Moon? Or how much power is necessary to boost rockets carrying building materials into space?

The solution may be near-Earth asteroids: NEAs have materials that can be processed to produce such structures, some with enough necessary materials to last for decades.

How do we know these bodies are worth exploring? Walk into the stretch of land called the Sudbury Basin in Canada, and you will notice bare rock outcrops and a scruffy forest, barely supported by a thin soil layer. But from space, satellites show an indistinct oval shape, originally circular and about 8 kilometers deep, deformed by 2 billion years of crustal movement and glacial, fluvial, and eolian erosion. Robert S. Dietz, an emeritus professor of geology at Arizona State University, was one of the first to establish the origin of the basin: an astrobleme ("starwound"), an eroded terrestrial impact scar so old that it no longer resembles a crater. Look close enough and some evidence does survive.

The object that struck was about 6.5 kilometers in diameter, in now central Ontario, just above the northern shores of Lake Huron. Today at Sudbury, the machines are noisy, digging for nickel and other economically important minerals. This is not only an example of space donating one of its treasures for the good of humanity; it is an example of using the resources of an asteroid that has hit the Earth.

Two types of asteroids seem to be the prime targets: carbonaceous chondrites containing carbon, primitive organics, and water and other volatiles, and the metallic type, with the platinum group and gold as possible by-products of refining the asteroids. The differences do not end there. Based on meteorite samples found on the Earth and spectrographic analysis of main-belt asteroids, the rocks are further broken down: carbonaceous chondrites and metallic rocks are joined by stony-irons, where silicate rock is peppered with chunks of metal; bright asteroids made of silicate and chondrite combinations; reddish asteroids (perhaps "rusted" rock of iron-containing materials or certain chemical combinations); and strange combinations we have yet to discover.

Mining the surface of an asteroid would be valuable in terms of resources, not only for space, but also for those left on Earth. Because of its low gravity, material could be easily removed from the asteroid, then sent home on special spacecraft built to carry large payloads. Some scientists even suggest using the "mass driver" technique (used previously to destroy planets in science fiction) to chew off pieces of the asteroid, and fling them into space toward the Earth, although no one mentions how to stop the projectile.

Differences in the asteroidal compositions lead many scientists to believe that we can use these diverse bodies as resources for future space endeavors. And if you add all the compositions together, asteroids even outshine the Moon in terms of resources. In fact, John Lewis, of the University of Arizona Space Engineering Research Center, estimates that the typical NEA may contain trillions of dollars (by today's standards) worth of metals.

Take for example, the differentiated, metal-cored asteroids. Millions of years of slamming bodies together have exposed asteroids rich with iron, chromium, nickel, and other precious metals. Such asteroid ores could be processed and used to build infrastructures for Earth- or Mars-orbiting space stations or materials for planet-based space colonies. Even by-products from refining the ores would be useful, including such metals as those from the platinum group. After all, it is estimated that a 1-kilometer metal asteroid contains about 8 billion tons of metal. Such an asteroid could supply the Earth with iron for 15 years, copper for 10 years, nickel for 1250 years, and cobalt for 3000 years.

WATER, WATER, BUT NOT EVERYWHERE

In order for humans to exist in space, there will be three obvious necessities: Shelter, food, and water. These three items will be at a premium in space. Shelter can come from a spacecraft; food can be carried or grown on an orbiting space station. But water will be the biggest problem, you cannot squeeze it from space or the spacecraft and it cannot be grown. In the wild imaginations of science fiction writers may come the truth: Water will be the future "money" of space travel. Whoever controls water will control space.

The biggest problem is water's weight. Shelters can be pushed into orbit with relative ease, and future metals and ceramics may make it that much easier to boost spaceships beyond the escape velocity of Earth. Food can also be packaged in lightweight wrappings or containers, and seeds can accompany space travelers into orbit (although the soils in which to grow the seeds may weigh significant kilograms, something soil scientists are working on).

Unfortunately, water is water, forever about 4 pounds per gallon. A potable water supply must be carried into orbit; if a space station or colony is being built, water will be needed for various uses.

It is not only getting the water into orbit; there are also the uses of water once you are in space. Probably more than anything, water can be used to make propellant—rocket fuel is hydrogen and oxygen, the elements of water—and once you have access to propellant, you can go virtually anywhere in the solar system, which is where the asteroids come in.

Many researchers believe that certain asteroids are dead comets, most of their volatiles burned out into the voids of space by constant runs around the Sun. The result would be a burned-out comet in a short-period orbit around the Sun, similar to the orbit of a near-Earth asteroid. Add to that the black, dusty, fine-grained material that will be coating the comet, and it is easy to see why the dead comets mimic asteroids.

If these dead comets do exist as near-Earth asteroids, they would be excellent sources of volatiles. Just brush aside the black coating of material from the surface of the burned-out comet and you will find plenty of the leftovers, including water and useful hydrocarbons. (Eugene Shoemaker gives a factor-of-two type estimate for the proportion of near-Earth asteroids that might be extinct comet cores; others say 20 to 60 percent.)

In fact, the primordial-like carbonaceous chondritic asteroids will likely attract the most attention: Studies comparing carbonaceous chondrite meteorites retrieved on Earth and their space counterparts indicate that C-type asteroids may hold up to 15 percent water, along with organic chemicals such as hydrocarbons and primitive organics, and other volatile elements.

Although asteroids may not be as water-rich as a comet (Comet Halley in its march around the Sun sent out more than 30 million tons of water vapor into space in a 6-month period), such bodies will still be perfect for providing water and fuel for trips in space. Water could be extracted from the hydrate rock mined from the asteroid. Lighter gases such as hydrogen and helium could be processed as fuel for certain spacecraft, such as those run by nuclear power.

Is it truly possible to find water in space in order to survive? There is a good chance, but in addition to looking at the asteroids, we may have to turn to other places in the solar system. There are even places that seem to be unlikely candidates for holding water, including Mercury (although it has not been confirmed), water on the Martian moon Deimos (which is estimated to be about 5 percent water), and water-ice in one of the Martian polar caps.

Ultimately, if enough water-ice deposits could be found on a near-Earth asteroid, the first steps in the human exploration of other planets would be greatly simplified. By the time we are ready to live off of the water in space, the technology may allow us to tow an asteroid, complete with water-ice, to a needy space colony. Special heat-generating tools could dig—actually, melt—their way through frozen mud, extracting water as they went. For the more claylike asteroids, water could be extracted by heating the clay in a cooker, the water and organics collected for future use. Water could be shipped from place to place with huge "super-tanker" spacecraft, with little problem loading the payload of valuable water because of the low-gravity environment. The water could then be sent to refineries to crack apart the hydrogen and oxygen, perfect for fueling spacecraft—a virtual "gas station" in space.

LIVING SPACE

How long each asteroid will last as a resource in space will depend on the asteroid type and on what resources are extracted. Logically, the larger ones may last for decades; smaller asteroids will last less time.

But what can we do with the leftover shell of an asteroid? Because of the intense radiation in space, such hollowed-out chunks of rock could be used for shielding space travelers from ever-present cosmic radiation and the occasional intense solar flare. Even in the beginnings of extrapolating materials from the asteroid, the empty space could be used by the miners for temporary respites from the constant bombardment of cosmic rays. Much smaller asteroids could also be broken down and used for ballast in spaceships or as counterweights for transportation vessels.

Building an actual mining colony on an asteroid has been dreamed of for years. For example, the 1989 winner of the National Space Society's Space Habitat Design Competition is typical: The Asteroid Resource Colony, developed by New York architect

Claudio Veliz and supported by a team including Raul Rosas, Wing Kin Lee, and John O'Connell. The colony, to support 18,000 people, would include a central mining hub around which four tethered habitation modules reside. The initial contact would stop the spinning of an Apollo asteroid; the hub would be built next. The habitation modules would then be built, rotating around the center hub to simulate about one-fifth the Earth's gravity.

Are we destined someday to live in an asteroid? Or will we one day consider the small bodies valuable resources, and not an enemy sneaking up on us? Such assets in space are hard to deny. As NEAR and future asteroid rendezvous and flyby craft will teach us, it may be easier than we think to "catch a falling star," in this case a passing "star," and put it in our space resources pocket.

CAN WE BECOME THE WATCHERS?

FEELINGS OF FALSE SECURITY

Thankfully, this book is not an "after-the-fact" book. Yes, I explained the historical background behind impacts, and I have cited instances of close and closest encounters in our own time. But most of all, this book is a "before-the-fact" book: a warning that we should be aware of the potential dangers of asteroid collisions and an encouragement as to how the asteroids may one day be the new frontier we flock to on our way to colonize our solar system.

I have also emphasized the efforts of many individuals who made, and are making, exhaustive searches of the sky for potentially harmful or useful asteroids and comets. Given all the time spent on telescopes at the major observatories, the Congressional involvement in Spaceguard and other search projects, international efforts to track asteroids, and the endeavors of amateur astronomers, we should feel secure. But in reality, the effort is not consistent. It is simply a small number of people trying to search for small objects in a vast amount of sky, using a finite amount of telescope time.

Would it be too much to ask to increase the networks of asteroid search telescopes, provide more funding for programs, and even develop more innovative ways to search the sky for these small objects?

I am realistic enough to know that even with more people and telescope time, there is always the chance that we will miss a potentially destructive asteroid that makes its way toward the Earth. But with more coverage, such chances of a strike are decreased exponentially.

There is so much more we can do: We can continue to scout

already imaged sections of the sky for possible near-Earthers. We can commit more funding, people, and time to examine past and future telescopic images. We can also use scientific methods from other fields and apply them to asteroid hazards. For example, Richard Binzel of the Massachusetts Institute of Technology suggested at a conference at the United Nations in April 1995 a "near-Earth asteroid hazard index" that would give scientists a way to communicate to the public the collision probabilities and consequences of close encounters. The idea is similar to maps used by scientists in the field of natural hazards: Fluvial geomorphologists divide flood regions into zones that represent flooding hazard along a river. Those within the floodplain are labeled as having the most potential for flooding hazards; those up-bank from the river decrease in flooding potential. Binzel proposes that such a hazard index be made for the asteroids: Those with the most potential to be troublesome to Earth would be designated a higher number (in his example, the numbers range between 1 and 5), and thus, would be watched and tracked with greater care. The lower-number asteroids or those with less potential to damage the Earth, would also be tracked; a hazard index of 0 would mean no likelihood of collision. In addition, Binzel suggests an index of possible consequences, ranging from no consequence (no damage likely, and probable atmospheric dissipation) to global (possible disruption of the global ecosphere, with severe regional damage).

THE IMPACT OF POLITICS

Whose responsibility is it to keep an eye on potentially harmful asteroids and comets? I believe it is everyone's responsibility, worldwide.

I realize, too, that not every country has the facilities or money to contribute to such studies, and all countries are concerned with the immediate needs of their own people. Today's headlines are rampant with social problems, such as unemployment and downsizing; military problems and civil war; or economic problems, such as the balancing the government's budget.

But, as every Earth-based natural disaster helps us to realize, nature has no political boundaries. Its only promise is that everything will change. And one of those changes may come in the form of an oversized asteroid impact from space.

And what about the potentials of asteroids for the future of Earth? If we were to actually capture several of these bodies, or just gather the needed minerals from afar and cart them back to Earth, all countries would benefit. The asteroids would be the new frontiers to spark technology and provide valuable resources for Earth, literally a pot of gold at the end of a cosmic rainbow.

Or perhaps we should look at this from another angle, albeit

FIGURE 1. The question is not, is Earth the target for an asteroid strike, but when will it be struck and how big will the asteroid be? (Photo courtesy of NASA)

what some may call an anthropocentric one. We have an obligation to humanity to keep our species going. Is it possible that we are the only intelligent life-forms in the universe? And if so, does that mean we are obligated to keep this one group healthy in order to follow through with the universe's ultimate plans for us? And if we are not, isn't survival ingrained in us, driving us to keep our species around for as long as possible?

Those of us who watch the funding being cut will continue to watch and hope that improving technology will fill in the gap created by the dwindling capital. We will continue to ask for support to search for asteroids, allowing scientists to continue to keep up with potentially dangerous asteroid discoveries.

In the end, we have two choices: Near-Earth asteroids will be our nemeses, exploding in our atmosphere or crashing into the surface, causing problems we have never faced before. Or near-Earth asteroids—or any asteroids—will be our saviors in space. They will provide us with the much-needed resources to carry on throughout the solar system and beyond. Our population will not decrease, and with an increase comes a necessity, need, and desire to expand. Not only that, if we expand into space, and a disaster, asteroid or otherwise, were to befall the Earth in the future, it would not mean the end of the human race.

In the meantime, we will keep watching as best as we can. And we hope beyond hope that we will be lucky enough to spot the one that eventually sneaks up from behind (see figure 1).

APPENDIX

Table 1. Close Approaches by Asteroids
and Comets 1996–2020

Object	Date/month		CA Dist (AU)
1991 CS	1996 08	28.8	.0508
4179 Toutatis	1996 11	30.0	.0354
6037 1988 EG	1998 02	28.9	.0318
1991 RB	1998 09	18.5	.0401
1992 SK	1999 03	26.3	.0560
6489 Golevka	1999 06	2.8	.0500
4486 Mithra	2000 08	14.4	.0466
4660 Nereus	2002 01	22.5	.0290
1994 PM	2003 08	16.7	.0245
1990 OS	2003 11	11.5	.0250
6239 Minos	2004 02	2.9	.0564
4179 Toutatis	2004 09	29.6	.0104
1992 UY4	2005 08	8.4	.0402
2340 Hathor	2007 10	22.2	.0600
1989 UR	2007 11	26.4	.0406
4450 Pan	2008 02	19.9	.0408
1991 VH	2008 08	15.5	.0458
4179 Toutatis	2008 11	9.5	.0502
1994 CC	2009 06	10.2	.0167
Honda-Mrkos-Pajdusakova	2011 08	15.3	.0601
4179 Toutatis	2012 12	12.3	.0463
1988 TA	2013 05	8.1	.0546
2340 Hathor	2014 10	21.9	.0482
1566 Icarus	2015 06	16.7	.0538
1994 AW1	2015 17	15.3	.0577
5604 1992 FE	2017 02	24.4	.0336
1991 VG	2017 08	7.4	.0568
3122 Florence	2017 09	1.5	.0472
1989 UP	2017 11	4.2	.0471
1991 VG	2018 02	11.9	.0473
1990 MF	2020 07	23.9	.0546

Close approaches for asteroids and comets to the Earth through 2020. Only those objects coming within 0.06 AU (5.6 million miles) of the Earth have been included. Comet Honda-Mrkos-Pajdusakova is the only comet on this list. (Courtesy of Don Yeomans, Jet Propulsion Laboratory, January, 1996)

Table 2. Asteroid Records

Record	Asteroid	Comments
Brightest	Vesta	Only asteroid from main belt that is visible to the naked eye
Darkest	Arethusa	Lowest reflectivity
Largest	Ceres	Approximate diameter 1003 to 1040 kilometers
Smallest	Hathor	Approximate diameter 0.5 kilometers (smaller ones have been detected, but are not named)
Longest rotation	Glauke	1500 hours
Shortest rotation	Icarus	2 hours, 16 minutes
Longest revolution[a]	944 Hidalgo	Over 14 years (just beyond the orbit of Saturn)
Shortest revolution	Ra-Shalom	283 days (orbit inside the Earth's orbit)
First asteroid with satellite	Herculina	Diameter about 217 kilometers with a satellite of 50 kilometers
Closest near-Earth asteroid	1994 XM1	Closest approach to Earth at [101,367 kilometers], a [13-meter] asteroid
Smallest semimajor axis	1994 GL	Found by James Scotti and David Rabinowitz, the asteroid averages just 0.683 astronomical units from the Sun (102 million kilometers), in a loop that takes it between the orbits of Mercury and Earth (revolution is 206 days)

[a]If the object called 2060 Chiron, discovered in 1977, is really an asteroid, it could claim the record for the most distant asteroid (it is currently numbered as an asteroid, but is considered a very large comet by many); or if any of the Kuiper objects discovered in the past few years are determined to be asteroids, they also may win the record. Another contender is 5145 Pholus (1992 AD), which revolves around the Sun in 93 years, longer than Chiron (51 years). Discovered in January 1992, the object is also the reddest object ever imaged, which may be an indication of organics on its surface.

ENDNOTES

All science writers are a product of their technological environments, and the Internet and other communication services are now a part of our research life. Therefore, numerous endnotes were gleaned from the Internet via World Wide Web homepages (listed here as "http:" addresses) and from personal communications gathered mostly from e-mail (or from contacts for prior magazine articles I wrote). Also note that many World Wide Web homepages change over long and/or short periods. I regret any inconvenience caused by the changing of a listed homepage address since publication.

CHAPTER 1

1. L. A. Marschall, *The Supernova Story* (Plenum, New York, 1988), pp. 74–75; Patrick Moore, Garry Hunt, Iain Nicholson, and Peter Cattermole, *The Atlas of the Solar System* (Crescent, New York, 1990), p. 414.
2. H. Couper with N. Henbest, *New Worlds: In Search of the Planets* (Addison-Wesley, Massachusetts, 1985), p. 115.
3. C. J. Cunningham, Giuseppe Piazzi and the "missing planet," *Sky & Telescope* **84**, 274–275 (1992).
4. T. Dick, *Celestial Scenery* (E.C. & J. Biddle, Philadelphia, 1851), p. 127.
5. D. K. Yeomans, *Comets: A Chronological History of Observation, Science, Myth, and Folklore* (Wiley, New York, 1991), p. 149.
6. W. Sheehan, *Worlds in the Sky* (University of Arizona, Tucson, 1992), p. 106.
7. E. L. G. Bowell, Lowell Observatory, personal communication.

CHAPTER 2

1. C. A. Ronan, *The Natural History of the Universe* (Macmillan, New York, 1991), pp. 30–33.
2. *Science News* **147**, 372 (1995).
3. Keck telescope looks at the big bang, *Science News* **145**, 349 (1994).
4. P. Halpern, *The Cyclical Serpent* (Plenum, New York, 1995), pp. 96–103.
5. Further evidence of a youthful universe, *Science News* **148**, 166 (1995).
6. O. R. Norton, *Rocks from Space* (Mountain, Missoula, MT, 1994), pp. 332–336.
7. Meteorite hints at pounding of planets, *Science News* **148**, 199 (1995).

CHAPTER 3

1. T. Dick, *Celestial Scenery* (E.C. & J. Biddle, Philadelphia, 1851), p. 127.
2. J. K. Beatty and A. Chaikin, ed., *The New Solar System* (Cambridge, New York, 1990), p. 231.
3. O. R. Norton, *Rocks from Space* (Mountain, Missoula, MT, 1994), p. 343.
4. S. Mitton, ed., *The Cambridge Encyclopaedia of Astronomy* (Crown, New York, 1977), p. 238.

CHAPTER 4

1. T. Dick, *Celestial Scenery* (E.C. & J. Biddle, Philadelphia, 1851).
2. S. Ostro, personal communication.
3. W. J. Webster and K. J. Johnston, in *Asteroids II* (R. P. Binzel, T. Gehrels, and M. S. Matthews, eds.) (University of Arizona, Tucson, 1989), pp. 192–211.
4. W. Hartmann, personal communication, IUGG Conference.
5. D. Durda, Two by two they came, *Astronomy* **23**, 30–35 (1995).
6. R. S. Hudson and S. J. Ostro, Shape of asteroid 4769 Castalia

(1989 PB) from inversion of radar images, *Science* **263**, 940–943 (1994).

7. D. Durda, Two by two they came, *Astronomy* **23**, 30–35 (1995).
8. R. Cowen, Do asteroids come in pairs? *Science News* **147**, 191 (1995).

CHAPTER 5

1. O. R. Norton, *Rocks from Space* (Mountain, Missoula, MT, 1994), p. 15.
2. Two more atens, *Sky & Telescope* **79**, 251 (1990).
3. Miscellaneous sources were used for this information: O.R. Norton, *Rocks from Space* (Mountain, Missoula, MT, 1994), pp. 185–241; G. Abell, *Exploration of the Universe* (Holt, Rinehart, and Winston, New York, 1969), pp. 358–361; http://seds.lpl.arizona.edu/nineplanets/nineplanets/nineplanets.html
4. K. K. Yau, P. R. Weissman, and D. K. Yeomans, *Meteorics* (November 1994).
5. *Sky & Telescope* **81**, 383 (1991).

CHAPTER 6

1. An organic asteroid? *Sky & Telescope* **85**, 15 (1993).
2. Richard Binzel, Massachusetts Institute of Technology, personal communication.
3. Peter Thomas, Cornell University, personal communication.

CHAPTER 7

1. D. K. Yeomans, *Comets: A Chronological History of Observation, Science, Myth, and Folklore* (Wiley, New York, 1991), p. 166, 178–179; http://encke.jpl.nasa.gov (Comet Observations).
2. J. C. Brandt and R. D. Chapman, Rendezvous in space: The science of comets, *Mercury* **21**, 178–192 (1992). (This is also the

name of Brandt and Chapman's book, published by W.H. Freeman in 1992.)

3. C. Matthews, personal communication.
4. D. H. Levy, *Impact Jupiter* (Plenum, New York, 1995), p. 237.
5. W. Corliss, *Science Frontiers* **91**, 1 (1994).
6. J. K. Beatty and D. H. Levy, Crashes to ashes, *Sky & Telescope* **90**, 18–26 (1995).
7. J. E. Bortle, Comet splits and champion discoverers, *Sky & Telescope* **89**, 107–108 (1995).

CHAPTER 8

1. C. W. Tombaugh and P. Moore, *Out of the Darkness: The Planet Pluto* (Stackpole Books, Harrisburg, PA, 1980), pp. 25–27.
2. Ibid, pp. 142–143.
3. This subject has numerous sources, including personal communications with Clyde Tombaugh and David Levy; another source of the debate occurred in D. Green, J. Chambers, G. Williams, and D. Levy, Focal point, *Sky & Telescope* **88**, 6–9 (1994).
4. C. W. Tombaugh, Plates, Pluto, and Planets X, *Sky & Telescope* **81**, 360–361 (1991).
5. Gerard Kuiper, in *Astrophysics* (J. A. Hynek, ed.) (McGraw-Hill, New York, 1951), pp. 357–424.
6. *Science Frontiers* **90**, 1 (1993); *Science News* **147**, 293 (1995); *Sky & Telescope* **90**, 10 (1995).
7. Clyde Tombaugh, personal communication.

CHAPTER 9

1. R. Burnam, Arizona's meteor crater, *Earth* **1**, 50–58 (1991); notes taken at Meteor Crater, 1991, at the Asteroid, Comet, and Meteor Conference, Flagstaff, Arizona.
2. G. Foster, *The Meteor Crater Story* (Meteor Crater Enterprises, AZ, 1991); J. Paul Barringer, personal communication; J. K.

Davies, *Cosmic Impact* (St. Martin's, New York, 1986), pp. 18–22.

3. D. Steel, *Rogue Asteroids and Doomsday Comets* (John Wiley, New York, 1995), p. 91.

4. R. S. Deitz, Shatter cones in cryptoexplosion structures, *Journal of Geology* **67**, 502–503 (1979).

5. Meteorite crater under Chesapeake Bay? *Astronomy* **23**, 24–26 (1995).

6. R. S. Dietz, Are we mining an asteroid? *Earth* **1**, 36–41 (1991).

7. Randy Cygan, personal communication; P. Barnes-Svarney, Blast from the past, *Technology Review* **94**, 23 (1991); Abstracts from planetary impact events: Materials response to dynamic high pressure, IV International Conference on Advanced Materials, August 28–29, 1995, Cancun, Mexico.

8. Did the "Big One" pack a one-two punch? *Sky & Telescope* **84**, 8 (1992); Eugene Shoemaker also gave a lecture on this topic at the Asteroid, Comet, and Meteor Conference, 1991.

9. Dating the Manson crater: No link to Chicxulub, *Sky & Telescope* **87**, 12 (1994).

10. J. K. Beatty, "Secret" impacts revealed, *Sky & Telescope* **87**, 26–27 (1994).

11. J. Erickson, *Target Earth!* (TAB Books, Blue Ridge Summit, 1991), pp. 65–66.

12. R. A. Gallant, Journey to Tunguska, *Sky & Telescope* **87**, 38–43 (1994); http://maxwell.sfsu.edu/asp/asp.html (Astronomical Society of the Pacific).

13. National Science Foundation, PR 94-56, Reading the "fine print" of climate history in Greenland's ice, September 28, pp. 2–3 (1994).

14. P. H. Schultz and J. K. Beatty, Teardrops on the Pampas, *Sky & Telescope* **83**, 387–392 (1992).

15. Virgil Barnes, personal communication; V. Barnes, Origin of tektites, *The Texas Journal of Science* **41**, 5–33 (1977).

16. J. O'Keefe, personal communication.

17. P. Barnes-Svarney, Legendary impacts, *Final Frontier* **13**, 7–8 (1993); D. Steel, lecture given at the Asteroid, Comet, and Meteor Conference, 1992, Flagstaff, Arizona.

CHAPTER 10

1. L. W. Alvarez, W. Alvarez, F. Asaro, and H. V. Michel, Extraterrestrial cause for the Cretaceous–Tertiary extinction, *Science* **208**, 1095–1107 (1980).
2. J. K. Beatty, Killer crater in the Yucatan? *Sky & Telescope* **82**, 38–40 (1991).
3. Yucatan impact dated, *Sky & Telescope* **85**, 12–13 (1993).

CHAPTER 11

1. http://nssdc.gsfc.nasa.gov/planetary/clementine.html.
2. H. J. Melosh, IUGG Conference, July 1995, Boulder, Colorado.
3. A. Y. Glikson, Asteroid/comet mega-impacts may have triggered major episodes of crustal evolution, *EOS* **76**, 49, 54–55 (1995).
4. C. Sagan and C. Chyba, Endogenous production, exogenous delivery and impact-shock synthesis of organic molecules, *Nature* **355**, 125 (1992).
5. P. Barnes-Svarney, The year of the asteroid, *Final Frontier* **4**, 16–17 (1991).

CHAPTER 12

1. T. Gehrels, Collisions with comets and asteroids, *Scientific American* **274**, 54–59 (1996).
2. E. F. Helin and R. S. Dunbar, Search techniques for near-Earth asteroids, *Vistas in Astronomy* **33**, 21–37 (1984).
3. D. Steel, *Rogue Asteroids and Doomsday Comets* (John Wiley, New York, 1995), p. 30.
4. B. G. Marsden, Comet Swift–Tuttle: Does it threaten Earth? *Sky & Telescope* **85**, 16–19 (1993); D. Steel, *Rogue Asteroids and Doomsday Comets* (John Wiley, New York, 1995), pp. 7–12.

CHAPTER 13

1. C. Chapman and D. Morrison, Chances of dying from selected causes—Table, *Nature* **367**, 39 (1994).

2. D. Morrison, *The Spaceguard Survey: Report of the NASA International Near-Earth-Object Detection Workshop* (Government Printing Office, Washington, DC, January 1992), pp. 7–12.
3. D. A. Rothery, *Satellites of the Outer Planets: Worlds in Their Own Right* (Clarendon, Oxford, 1992), p. 59.
4. D. Morrison, Target Earth: It will happen, *Sky and Telescope* **79**, 261–265 (1990).
5. D. Morrison, *The Spaceguard Survey*, pp. 10–11.
6. M. W. Browne, Mathematicians say asteroid may hit Earth in a million years, *NY Times*, April 25, 1996.
7. http://ccf.arc.nasa.gove/sst/main.html and http://bozo.lpl.arizona.edu/nineplanets/asteroids.html

CHAPTER 14

1. Jay Gunter, personal communication; B. Gunter, personal communication.
2. Roger Harvey, personal communication.
3. Brian Warner, personal communication.
4. D. Morrison, Target: Earth! *Astronomy* **23**, 34–41 (1995).
5. Edward Bowell, personal communication.
6. http://www.halebopp.com; A. Hale, personal communication.
7. http://bozo.lpl.arizona.edu (Report of near-Earth objects survey).

CHAPTER 15

1. P. Barnes-Svarney, Staking a claim, *Ad Astra* **4**, 25–26 (December 1992).
2. P. Barnes-Svarney, The craft of CRAF, *Ad Astra* **3**, 18–21 (1991).
3. S. L. Murchie, A. F. Cheng, and A. G. Santo, Encounter with Eros: The near-Earth asteroid rendezvous mission, *Lunar and Planetary Information Bulletin* **75**, 2–5 (1995).
4. R. Farquhar and J. Veverka, Romancing the stone: The near-Earth asteroid rendezvous, *The Planetary Report* **15**, 8–11 (1995).
5. P. Barnes-Svarney, Grabbing a piece of the rock, *Ad Astra* **2**, 7–13 (1990).

SELECTED BIBLIOGRAPHY

Binzel, R. P., Gehrels, T., and Matthews, M. S., eds., *Asteroids II* (University of Arizona, Tucson, 1989).

Chapman, C. R. and Morrison, D., *Cosmic Catastrophes* (Plenum, New York, 1989).

Cunningham, C. J., *Introduction to Asteroids* (Willmann-Bell, Richmond, 1988).

Davies, J. K., *Cosmic Impact* (Fourth Estate, London, 1986).

Gehrels, T., ed. *Asteroids* (University of Arizona, Tucson, 1988).

Hodge, P., *Meteorite Craters and Impact Structures of the Earth* (Cambridge University, Cambridge, England, 1994).

Hutchinson, R. and Graham, A., *Meteorites* (Sterling, New York, 1993).

Mark, K., *Meteorite Craters* (University of Arizona, Tucson, 1987).

McSween, H. Y., Jr., *Stardust to Planets: A Geological Tour of the Solar System* (St. Martin's, New York, 1993).

Raup, D. M., *Extinction: Bad Genes or Bad Luck?* (W.W. Norton, New York, 1991).

Rothery, D. A., *Satellites of the Outer Planets: Worlds in Their Own Right* (Clarendon, Oxford, England, 1992).

Schmadel, L. D., *Dictionary of Minor Planet Names* (Springer-Verlag, New York, 1992).

Sheehan, W., *Worlds in the Sky: Planetary Discovery from Earliest Times through Voyager and Magellan* (University of Arizona, Tucson, 1992).

INDEX